KNIGHTS AND CASTLES

Knights and Castles

Minorities and Urban Regeneration

Edited by

FRANCESCO LO PICCOLO
University of Palermo

HUW THOMAS
Cardiff University

LONDON AND NEW YORK

First published 2003 by Ashgate Publishing

Reissued 2018 by Routledge
2 Park Square, Milton Park, Abingdon, Oxon OX14 4RN
711 Third Avenue, New York, NY 10017, USA

Routledge is an imprint of the Taylor & Francis Group, an informa business

Publisher's Note
The publisher has gone to great lengths to ensure the quality of this reprint but points out that some imperfections in the original copies may be apparent.

Disclaimer
The publisher has made every effort to trace copyright holders and welcomes correspondence from those they have been unable to contact.

A Library of Congress record exists under LC control number: 2002034492

ISBN 13: 978-1-138-70844-0 (hbk)
ISBN 13: 978-1-315-19863-7 (ebk)

Contents

List of Tables

List of Contributors

John Foley
University of New Orleans
New Orleans
USA

Elise Henu
Université d'Aix Marseille III
Marseille
France

Abdul Khakee
University of Örebro
Örebro
Sweden

Robert Kloosterman
University of Amsterdam
Amsterdam
The Netherlands

Björn Kullander
University of Örebro
Örebro
Sweden

Mickey Lauria
University of New Orleans
New Orleans
USA

Francesco Lo Piccolo
University of Palermo
Palermo
Italy

Silvia Macchi
Università "La Sapienza"
Rome
Italy

Richard Silburn
University of Nottingham
Nottingham
UK

Paola Somma
Independent scholar

Huw Thomas
Cardiff University
Cardiff
UK

Joanne van der Leun
Erasmus University
Rotterdam
The Netherlands

Acknowledgements

The forerunners of the chapters in this book were first presented as papers at an international workshop held at the International Centre for Planning Research, Cardiff University, in September 1999. Angela Evans worked miracles in producing a manuscript efficiently and with good humour; we are all in her debt. We are grateful to all who participated in the workshop for their ideas and comments, which helped shape this book. We are also grateful to the International Centre for hosting and assisting with the funding of the workshop. In addition to the authors of the chapters, the participants included Eid Ali Ahmed, Tom Blair, Edward Hulsbergen and Donna Liburd.

The title of the book alludes to a metaphor of Vittorio Foa, which he used to highlight different strategies for living. The Castle moves in a straight line, encountering conflict and struggle as part of an unavoidable pre-defined field of action; the Knight moves in a more oblique fashion, searching for different fields and levels of action and circumventing opposition.

Chapter 1

Regeneration, Minorities and Networks

HUW THOMAS

Introduction

This book considers positive roles of minorities in urban regeneration (a term interpreted liberally, for our purposes) through a series of case studies, largely within Europe.[1] Seeking case studies of urban renewal which can be viewed in some way positively from the perspective of minority groups is important, but also dangerous. One of the more obvious dangers is the possibility that drawing attention to positive case studies may suggest that minorities are generally rather sensitively considered in urban renewal in Western countries. Of course, this would fly in the face of considered evaluations of urban renewal (see, e.g., Khakee et al, 1999). Yet to focus exclusively on the systematic exclusion and poor treatment of minorities in urban renewal has its own dangers.

First, it can present too crude, and misleading, a picture of the agencies and workers engaged in urban regeneration. Whilst we must not ignore the very real effects of institutional discrimination,[2] and acknowledge the structural constraints on individual effort, we must not overlook the efforts of a number of workers to produce progressive results in difficult circumstances. This is not just a matter of fairness to them; they also represent a vital resource in the struggle to create a more just approach to urban renewal. It is important to try to analyze the circumstances in which their efforts seem to be particularly successful.

Secondly, focussing exclusively on the limitations and negative outcomes in relation to minorities can buttress popular stereotypes of them as (passive) victims and a social problem. This is a well-founded criticism in relation to the implicit portrayal of black and ethnic minorities in discussions of planning and urban policy, for example (see, e.g., Thomas and Ritzdorf, 1997; Thomas, 2000). In denying them agency, such accounts ignore the way in which minorities are active in ordering their lives, and may create spaces and institutions which support and instantiate values and perspectives more or less at variance with the "mainstream" (Herbst, 1994; Pile, 1999 pp. 30–34). (This is not to deny the importance of the imbalances of power in relation to which minority status is defined; simply to say that power is resisted, and this is important to highlight in the context of a continuing struggle to transform social relations.)

The adaptability and creativity of minorities is currently being challenged by the profound changes occurring (at varying speeds) in the governance of Western countries (see, e.g., Thomas and Lo Piccolo, 2000). In the "differentiated polity" which Rhodes (1997, p. 3) has described as emerging, formal and publicly defined relationships between agencies and individuals are supplemented by an array of more fluid relations, often characterized as "networking". These changes present both threats and opportunities to groups currently excluded from significant influence in governance. On the one hand the erosion of institutional relations which have a history of exclusion at least offers minorities a new "game" in which they may achieve a little more success. On the other hand, the rules of the new game are considerably less well defined than those of the old. Networking is, in essence, a (generally covert or low profile) exclusionary activity. It has, of course, been an important way of excluding minorities even within the formalities of representative democracies; but the formalities and bureaucracies of traditional forms of representative government did allow "outsiders" to secure policy gains on occasion if they organized well – sometimes with spectacular results as in the case of the Trotskyist Militant group which controlled local government in Liverpool in the 1980s (for rather different examples see Clavel, 1986). The fluidity of contemporary governance makes it even more difficult for excluded groups to appreciate how it is they are being excluded, and hence to counter it. Reviewing positive stories may offer some pointers as to how these difficulties may be overcome.

For these advantages to be achieved, however, an analytical framework is needed within which case studies can be considered systematically. The ambition of the book, therefore, is to collect positive stories, but to suggest, also, an analytical framework within which they may yield some more enduring insights into creating conditions in which minorities benefit from and influence urban regeneration. The framework, which is sketched, has two main elements. First, that minorities are socially constructed, a process within which urban regeneration activity itself plays a part; moreover, their emergence as socially (including politically) salient categories is temporally and spatially variable, yet some systematic understanding of this variability is possible. Secondly, where particular minority categories do emerge as identities around which people mobilize in urban regeneration, then it is the permeability of key networks, and the capacity of minorities to develop and exploit new networks (of various kinds) which is central to their involvement in, and benefiting from, urban regeneration. The remainder of this chapter elaborates a little on these points.

Minorities

To say that minorities are socially constructed is to deny that minority status – be it related to disability, race or ethnicity, for example – denotes some essential, unchanging characteristic. It is to say that the classification of minorities – the boundary creation, and associated markers, for example – is a social process, subject to temporal and spatial variation, and reflecting social relations (notably, relations of power at a particular time). Denotations such as "race" are in Hall's (1996) phrase "floating signifiers", the precise referents for which, at any time need to be investigated empirically. The process of minority-formation is a socio-spatial process, for as Thomas et al (1996) have argued social identities are, typically, bound up with entitlements to be in, or have use of, certain kinds of places. To be a certain kind of person is, in part, to have the right (de facto, if not de jure) to be in a particular kind of place (racial segregation is perhaps the crudest illustration of these processes).

Such examples remind us not only that minority status is socially constructed, is contestable and bound up with relations of power (actual and fantasized – Moore, 1994) but that urban regeneration, neighborhood revitalization, and the like are arenas in which these processes occur. Given that the construction of ethnic categories, for example, involves defining "insiders" and "outsiders", and such definitions are suffused with relations of power, then it follows that when ethnicity seems a particularly potent element in urban regeneration (e.g. Carmon's (1998) account of the seemingly positive role of Jewish immigrants in revitalizing parts of Israeli cities) we might ask what the implications of this were not only for one ethnic group, but also for its relations with those outside the group (in Carmon's case study referred to above, for example, what were the implications of settlement leading to urban regeneration by recent immigrants to Israel for Palestinian residents of the city?). A further set of pertinent critical questions might address relations of power within minority ethnic groups. Afshar (1989) argues that the economic significance of kinship networks among Pakistani immigrants to a UK city reinforced (and depended upon) conventional gender roles which left women with a restricted sphere within which to exercise choices about how they lived their lives (cf Levine (2001, p. 71) on gendered activities among immigrant German Jews in New York State). The history of the Tupperware party (Pile, 1999) is one of economic activity being promoted within existing gender roles; the attraction of Tupperware was that it allowed, indeed required, women to develop social networks, but these did not challenge gender relations (including power relations bound up with hegemonic gender roles). On the other hand, having an independent source of income itself empowered women. A delicate balance in power relations within families was therefore constantly being maintained. We see similar processes in accounts of entrepreneurial women in both Britain and France who have exploited knowledge and social networks to which they have access as members of minority ethnic groups. Hardill and Raghuram's (1998) case studies of "Asian women in business" in the UK explored, inter alia, the complexity of the connections, or networks, in which female Asian entrepreneurs operated – it emerged that they had significant overlapping social and business links both within and outside the UK. The ability to develop, or exploit, these networks depends on a supporting technological and social infrastructure (in the sense of supportive norms), but it also begins to re-fashion them. Thus Body-Gendrot (1999) has no illusions about the social, economic and political constraints facing ethnic minority women who develop businesses catering for the distinctive needs of women of their ethnic group. However, she also points out that some women become skilled at bridging between minority networks and governmental bureaucracies (for example, in order to gain grant aid for community economic projects); this in itself may be a source of influence *within* a minority group. Certainly this is what Ui (1991) reports in relation to Cambodian immigrants in California, where employment opportunities available to women, largely related to their location in an ethnic enclave, helped re-shape domestic gender roles.

The very notion of minority, then, needs to be engaged critically for a full analysis of "what is going on" in a particular case study – in this book, Macchi in relation to redevelopment and regeneration in Rome, and Henu in a French case study interrogate the ways in which ideas of "mental illness" and "ethnicity" respectively are constructed (and sometimes undermined) in grass-roots mobilization. In Macchi's case study, for example, the struggle to secure a constructive community use for a psychiatric hospital led to a questioning of a distinction between patient and health professional which had, hitherto, been sustained by a particular institutional form, and its embedded power-relations (a workplace regime, one might say, but of course, one which had links with – and drew legitimacy from – power saturated distinctions extending far beyond the workplace). For our purposes in this book (as Somma argues in her chapter) inequalities of power are central to the construction of, and acquisition of, minority status – often the notion of dependence (see, e.g., Oliver, 1993) seems central to creation of a minority. Foley and Lauria's case study of New Orleans' French Quarter in this book reminds us that power must be understood relationally and, consequently, spatially. This means that the label "minority" cannot be regarded as a fixed tag which picks out essential characteristics – the (largely white) residents of the French Quarter may be relatively powerless within the narrow confines of the city's domestic politics, but the spatial geometry of power is complex and many will be enjoying privileged social relations which extend beyond the city's boundaries.

But though this broad theoretical context is vital for rigorous comparative study of mobilization in urban regeneration there remains a need for the identification of middle-range concepts which can bridge the gap between general theoretical perspectives and concrete actions in particular case studies. In an earlier publication, Khakee et al (1999) argued for the value of the idea of policy processes as such a middle-range concept, but while this idea must play a role in understanding aspects of exclusion within governance (see, e.g., Davoudi and Healey, 1995) it appears to be too restricted a notion to help explain minority involvement in urban regeneration as a whole – an activity which involves processes of governance, but spills over into civil society and the economy. A more hopeful candidate, however, has also found renewed life recently in the governance literature.

Networks

Rhodes (1999) has suggested that it is the increasing significance of networking which characterizes emerging forms of governance, at all spatial scales, but the notion of networking has a long history in the analysis of social behavior, for in Thompson et al's (1991) phrase, it is one of very few distinctive approaches to analyzing "the co-ordination of social life". In talking of co-ordination we must always beware of lapsing into an unreflective functionalism; but while acknowledging the frictions, inefficiencies and conflicts of everyday social activities (be they economic, governmental or recreational, for example) we need also to recognize that, generally, these occur against a background of social life which is, for lack of a better word, co-ordinate. Janet Foster (1997) has argued that even in impoverished housing areas popularly stigmatized as sites of alienation, despair and social (often radicalized) conflict, there are important social networks which sustain (and, of course, help define the boundaries of) everyday existence. Campbell's (1993) account of British housing estates characterized by disorder shows that networks of women, in particular, ensured that the physical necessities of everyday life were secured, and indeed, formed a basic for modest initiatives (not always successful) to improve physical, social and economic conditions in the area. The growth of networking in governance – not confined to Britain – may well be significant in helping to understand the involvement of minorities in regeneration. The notion of networking, however, has been employed to help understand a variety of social phenomena beyond governance. Labor market segmentation, for example, is typically underpinned by exclusionary networks (as Kloosterman and Van der Leun mention in relation to their Dutch case study in this volume). Sometimes, such exclusions which labor market networks deploy revolve around ethnic identity (Allen et al, 1998). More positively, Levine (2001, p. 7) has argued that for immigrant German Jews who fled Nazi Germany "extended kin and ethnic networks were a necessary component in enabling the refugees to capitalize on the structure of opportunity that awaited them in south-central New York".

The book will allow an evaluation of the usefulness of the concept of networking as an organizing idea for considering diverse aspects of urban regeneration. Urban regeneration is a process in which the state is centrally involved in promoting a physical, economic and (often) social transformation of an area, often (perhaps, typically) in concert with business and/or not-for-profit agencies of various kinds. It does seem possible to characterize this process (largely, not wholly) in terms of networks and networking. As Rhodes (1999) and others have argued (see Stoker, 1999) networking is central to contemporary governance and, hence, we might say, the *promotion and management* of regeneration. On occasion, minority groups can influence these processes by finding a role in appropriate networks (Nanton, 1998), though as Skelcher et al (1996, pp. 25–27) put it, there are significant filters which can operate to exclude minorities, even in ostensibly community–based regeneration. The *objectives* of regeneration might also be characterized in terms of the establishment of, or underpinning of, certain kinds of social and economic networks, and, perhaps, the disruption or eradication of others. Few planners today would be as open as Wilfred Burns in his assessment of the effects of 1960s urban renewal on socially marginal communities,[3] but there can be little doubt that regeneration activities do involve the promotion of particular kinds of ways of life, with the social and economic networks that these entail. Current concerns about new mixed tenure (and mixed class) housing developments in British inner cities (so-called, urban villages), for example, revolve around the possibility that there may be inadequate networking between socially diverse groups (Urban Task Force, 1999).

Within the governance literature doubts have been expressed about the explanatory power of the idea of networks (Dowding, 1995; John, 1998, pp. 85–91). The argument goes that the idea of a network may describe a pattern of social interaction, but that pattern itself requires explanation, and – once the further, more basic explanation is given – the notion of a network will become redundant. Such an argument is a powerful reminder of the limitations of an explanation of social life couched exclusively in terms of networking, but it does not follow that the notion of networks must be abandoned. Of course, explanations of who is included and excluded from networks, why some networks are influential and others not, and so on, must have reference to a range of other sociological concepts, but the idea of a network may well play an important part in an explanation of how certain activities are undertaken, and results achieved. There is a difference – and a difference which is important for public policy – between the kinds of exclusionary strategies which are used in, say, hierarchies as opposed to networks. Now, in both cases the effect may be to bolster, say, class inequalities, but the way in which this is achieved, and the aspect of class differences which is exploited, may well be different in the two cases. From a public policy perspective, that can be important. Consequently, in this book we explore aspects of networking, while actually conscious that the notion itself requires – in any particular case – to be bedded into a broader explanatory framework.

What, then do we mean by "networks", "networking" and related terms? Rhodes (1999) has recently characterized networks in governance as relationships based on "complex resource *exchanges* and *interdependencies*" (p. xviii), where mutual trust is an essential pre-requisite for network maintenance, *diplomacy* is the typical means of resolving conflict (as opposed to invoking authority, for example), and networks are characterized by a culture of *reciprocity* (all emphases added). Though Rhodes is concerned with networks in governance, I would suggest that the characterization is applicable to networks in other areas of social life (see for example, Thompson et al, 1991). There will be differences, however, in the significance of individual aspects of the characterization: for example, resolving conflict may be more important in policy networks than within the labor market, where disputes may, generally, simply mean that a person, or persons, leave the network. Of course, I am not suggesting some a priori principle, here – the importance of particular characteristics will be a contingent matter. An important question in any discussion is whether there are certain socio/economic, or political, conditions which are particularly supportive of networks. It is plausible to speculate that shared minority status – e.g. claiming a certain ethnicity – may in certain circumstances help a group of people or organizations develop the characteristics set out above; indeed, that to an extent constructing a minority ethnic identity may be bound up with establishing, or becoming part of, distinctive social networks. This book will provide some evidence, which will help evaluate that claim in relation to urban regeneration, with a particular focus on how minority status may be constructed, and used, within networking.

The literature on networking might lead us to expect that it plays an important role in shaping urban regeneration activity. Powell (1991), for example, argues that networking is particularly suited to the exchanges of resources, which are difficult to commodify, and he identifies qualitative matters, such as know-how, as an example of such a resource. The significance of know-how (of knowing "the system", and so on) in order to influence, and benefit from planning and urban regeneration, hardly needs underlining (Healey et al 1988, Thomas and Imrie, 1989). Know-how is a sensitive resource to exchange, and is one example which highlights the significance of trust in sustaining networks.[4] Frances et al (1991, p. 15) argue that analyzing how trust can be generated and retained "remains the prime problem in understanding networks", though Lorenz (1991) qualifies this by arguing (on the basis of case studies of business networks) that networking does not involve blind trust; firms trust where they must but are constantly seeking to minimize the risk thereby created. This resonates with Richard Silburn's observations in chapter eight about the significance of trust in creating effective partnership networks in regeneration, even in a civic world where "shallow partnership" is a more or less unavoidable stage in pursuing regeneration.

Case studies such as those of Werbner (1991) and Hardill and Raghuram (1998) point to the potential use of perceived shared cultural backgrounds – cemented for example, by the exchange of symbolically significant goods – in developing a basis for trust, and Lorenz (1991) underlines the importance of experience in shaping trusting behavior. It is a plausible suggestion that shared experience of minority status may provide a basis for trust. In addition Herbst (1994) has argued that exclusion from mainstream networks may help define some groups as minorities and in certain circumstances induce them to develop alternative networks (what she terms "parallel public spheres") connected to mainstream networks by "back channels" and brokers (Hannerz, 1980, pp. 181–201). There is some evidence of this phenomenon in urban regeneration processes in Britain (Thomas et al, 1996; Nanton, 1998), which raises important questions about how individuals (typically members of minorities) become identified as the links between mainstream and alternative networks, and what the consequences are for their standing in, and commitment to, minority groups (Cain and Yuval-Davies, 1990; chapter eight of this book).

Shared experience associated with minority status may also help shape the nature of the resource, which can be exchanged. In governance networks, for example, minorities, typically, will have influence to the extent that they can organize themselves to act cohesively. So, for example, the dense, and stable, social network of Bhattra Sikhs in Cardiff appears to have been a basis for the community's gaining a toehold in governance networks, as it has allowed a degree of cohesiveness and unity in the way Sikhs relate to the local political order (Ghuman, 1980). Here we see networking operating in two different ways – as the basis for an exchangeable resource, and as the mode of governance within which that resource can be exchanged. The Cardiff example illustrates too, the potential of the network idea to explore, in a systematic way, the nature or relations within it (their density, their degree of multiplexity) and to identify key roles (which might correspond to individuals or institutions/ practices) (Hannerz, 1980, pp. 181–201). On the other hand, Silburn (chapter eight) suggests that ethnic networks (which are spatially extensive) may undermine identification with residential neighborhood on occasion, and to that extent may *not* assist urban regeneration.

Rhodes says nothing about power in networks, but it is clear – *pace* Frances et al (1991) – that those participating in a network are not equal in power. Networks may not be formal hierarchies, but power flows through them nevertheless. A (relative) lack of exchangeable resources, for example, may leave some network members on the periphery; while skills such as diplomacy are culturally-embedded, and people of certain class-backgrounds, say, may find it difficult to develop the appropriate ones. Skelcher et al (1996) talk of a series of "filters" – individual, institutional and societal – which serve as exclusionary devices for networks involved in urban regeneration. It is clear that these filters will tend to sieve out those of minority status, as we have defined the term. But there can be circumstances in which this status forms the basis for access to networks, and this book will examine some examples of that kind. On other occasions, minorities may survive the filters, only to be marginal members of a network (Brownill et al, 1999).

Yet, a degree of authority within networks may sometimes be necessary for the network to function, or perhaps, survive. It may, therefore, be appropriate in some cases to talk of "network leadership". Leadership, in this context, needs to be considered as a capacity; the key question is, to what extent (and how) can networks develop informal or formal mechanisms to secure coherence and purpose in network activity (Judd and Parkinson, 1990). The example of Bhattra Sikhs referred to earlier suggests that ethnicity may be a useful underpinning for such a capacity; but of course, the capacity may be developed at the price of reinforcing inequalities of influence and power *within* the ethnic groups (see McDowell, 1999). It was a reluctance to countenance the emergence of powerful leaders which was one of the inhibiting factors in the development of a politically influential minority ethnic identity in Bolton in the 1970s (Hahlo, 1998). Riley (1999) examined some aspects of leadership in a case study of a network of organizations, but her key points, I suggest, can be applied to leadership in any network. She found that leadership in networks relied not upon formal authority, but upon legitimacy that needed to be earned and constantly renewed. To an extent, this legitimacy derived from personal style and ability. Nevertheless, strength (in terms of resources, and, indeed, standing outside the network) was also significant in bolstering legitimacy. Following this line of argument, we might speculate that gender relations outside a particular network might nevertheless have an impact on issues of leadership within the network. More generally, regeneration, and its attendant networking, takes place within a set of economic circumstances and public policies which must largely be taken as given by any particular locality.

Even as schematic an account as this provides us with a set of issues which the case studies in this volume can begin to explore. First, they may shed some light on the extent to which minority identity may prove to be useful (or essential) for the emergence of various kinds of networks within particular localities. Secondly, the case studies can help us examine some, or all, of the key characteristics of networks, for example leadership/legitimacy; dispute resolution; and resource exchange, and the ways these interact with a wider socio-political and economic context. Taken together, addressing these issues will provide fresh insights into the nature of contemporary urban regeneration as well as providing specific lessons or guidance in relation to increasing the influence of minorities within regeneration processes.

The Book

The chapters in the book were first presented at a workshop at Cardiff University in late 1999. This was an exploratory event, an opportunity for researchers and people with operational experience of urban regeneration to consider issues raised by the attempt to define successful outcomes of urban regeneration for minority ethnic communities. A wide variety of case studies were presented, in part reflecting the broadness of the terms "urban regeneration" or "urban renewal". Moreover, participants had a variety of reservations about key terms, such as "minority", and about the significance of the strategy of trying to identify success in the terms proposed, as was pointed out earlier in the chapter. Consequently, the chapters which follow do not present a development of a program so much as reflections on a broad theme, out of which, it has been argued above some useful strands for future understanding can be pulled.

Chapters two and three (by Henu and Macchi respectively) present case studies of socio-spatial change within a city where a group consciousness, specifically a consciousness of being a minority, was an important part of the mobilization against redevelopment proposals perceived to be threatening. Henu's comparative approach, in considering proposals to redevelop/renew two areas of Marseilles, allows her to portray with some conviction the differences between modes of mobilization in the two areas, differences which can only be properly understood in the light of their different histories – histories which explain the demographic composition of the area but also a history of relationships with outside agencies, particularly state agencies. The reaction of residents in the two areas – and the nature of any solidarity and of tensions – were influenced by these specific histories. Similarly, Macchi's approach is to introduce the reader to the (recent) historical context for struggles over the future of the site of a largely redundant psychiatric hospital in Rome; specific episodes can only be understood against this background. Macchi also highlights the way in which the struggle over a particular site was a process in which social relations, and the identities formed within them, were questioned and reconstituted: specifically, notions of mental illness and the sharp distinction between the sick and carers were broken down. It is clear from Macchi's accounts that particular individuals and groups were instrumental at given times in promoting change, and Henu, too, notes the significance of a group of residents and of state workers who networked with each other in order to produce particular desired outcomes (Henu is by no means convinced that this is altogether a good thing).

Kloosterman and Van der Leun (chapter four) share some of the themes of Henu and Macchi, but allude to them in a very different policy context. Their concern is to explore the extent to which different cities offer different kinds of economic opportunities to ethnic minorities, and the extent to which these are reflected in different economic outcomes. In their hands, the comparative approach is again a valuable tool for exploring spatial variation, in this case in the opportunity structures of Amsterdam and Rotterdam. These structures, and responses to them, require an historical analysis; moreover, it is clear that ethnic networks play some role in allowing immigrants to the Netherlands to find employment within the opportunity structures presented to them.

In chapter five Foley and Lauria provide a subtle discussion of the complexities of the politics of preservation and boosterism in the historic French Quarter of New Orleans. The contested meanings of place are explored within a narrative alive to the significance of spatial scale in the analysis of power, and there are sensitive interpretations of interview data with black, white and gay residents. It is clear that there can be no single uncontested answer to the call for positive outcomes for minorities; and their case emphasize that what is needed is to evaluate outcomes according to key principles not by reference simply to the outcome for a group labeled "minority".

Palermo, Sicily, has had minority ethnic communities within it for as long as anywhere in Europe and Sicilian culture is a wondrous hybrid of various influences. Yet the contemporary political challenge to ameliorate the conditions of minorities who are among the city's poorest residents is as great here as anywhere. Lo Piccolo, in chapter six, evaluates attempts by the current mayor to build a political coalition, which will support positive action to provide facilities and other assistance to immigrants.

The nature of the welfare state and the organization of civil society in Italy is completely different to that in Sweden; yet in both cases the importance of minorities gaining access to social and political networks is clear. In chapter seven Khakee and Kullander illustrate the barriers which deter ethnic minorities from becoming involved in tenants organizations on Swedish housing estates. Nevertheless they conclude that the way forward is to work with the grain of minority ethnic networks, networks developed on the basis of experiences of shared adversity and ascription of ethnic identity by the state and civil society.

In chapter eight however, Silburn explores the complexities of "working with the grain". His sensitive analysis of a multi-ethnic neighborhood of Nottingham exposes tensions within, and between, ethnic and radicalized groups. It is clear that the significance of involvement in governance networks and partnerships varies between ethnic communities at any given time, and indeed, varies over time. In a sense then, there is no simple "grain" to work with; and the activities which constitute promoting regeneration – including the creation of partnerships – and the products of regeneration itself, will influence the ever developing, ever changing nature of the "grain" of community life, and their networks.

Somma, in chapter nine, is less phlegmatic, and more suspicious of a focus on ethnic identity. Perhaps following what in France would be regarded as the republican tradition she wishes to fight racism, but is suspicious of placing too much weight, or hope, on the potential of ethnic identity as a mobilizing force for pursuing social justice. Racism is opposed as a violation of an individual's rights, not an injury to an ethnic group. Focusing on the latter runs the risk, as we noted earlier, of obscuring inequalities and oppression on grounds other than ethnicity, e.g. gender. It is fitting that an exploratory book should contain a thoroughly skeptical chapter. Absorbing some of its strictures is one of the lessons to be learnt from the book. But there are others, and these are reviewed in the book's final chapter, co-authored by the editors (and organizers of the workshop, which gave rise to the book).

Notes

[1]   "Urban regeneration" is a term devalued by its being caught up in city boosterism. Our case studies tend to be about physical transformation of cities associated with attempts to recast dramatically social and economic relations. The referent of the term "minorities" is context-dependent. It is a term used to identify those with little power and influence in particular circumstances, where those groups also constitute a numerical minority (though women are often referred to as a minority in western countries where they typically constitute a numerical majority of the population). Relations of power are what we are interested in, in our book, so for our purpose billionaires, say, do not constitute a minority.

[2]   Macpherson defines institutional discrimination (1999, p. 28) as: the "collective failure of an organization to provide an appropriate and professional service to people because of their color, culture, or ethnic origin. It can be seen or detected in processes, attitudes and behavior amounting to discrimination through unwitting stereotyping which may disadvantage minority ethnic people." Clearly the principle can be applied to analyze the discrimination against any minority.

3   Ward (1994, p. 155) quotes Burns, an influential planner of the 1960s, as identifying one of the tasks of urban regeneration as follows: "The task, surely, is to break up such groupings even though the people seem to be satisfied with their miserable environment and seem to enjoy an extrovert social life in their own locality."

4   Some kinds of know-how can, of course be commodified, but the episodic, sometimes intense, often unpredictable process of "hand-holding" through complex social or governmental systems is not something commercial consultancies relish undertaking. Moreover, for many people access to commodified know-how is impossible for financial reasons, so networking is essential.

References

Afshar, H. (1989), "Gender roles and the 'moral economy of kin' among Pakistani Women in West Yorkshire", *New Community*, 15(2), pp. 211–226.

Allen, J. et al (1998), *Rethinking the Region*, London: Routledge.

Body-Gendrot, S. (1999), "Pioneering Moslem women in France", in Brook, C. and Pain, K. (eds), *City Themes*, Milton Keynes: Open University.

Brownill, S. et al (1999), "Patterns of inclusion and exclusion: ethnic minorities and Urban Development Corporations", in Stoker, G. (ed.), *The New Politics of British Local Governance*, London: Macmillan.

Cain, H. and Yuval-Davies, N. (1990), "The equal opportunities community and the anti-racist struggle", *Critical Social Policy*, 10(2), pp. 5–26.

Campbell, B. (1993), *Goliath: Britain's Dangerous Places*, London: Methuen.

Carmon, N. (1998), "Immigrants as carriers of urban regeneration: international evidence and an Israeli case study", *International Planning Studies*, 3(2), pp. 207–225.

Clavel, P. (1986), *The Progressive City*, New Brunswick, NJ: Rutgers University Press.

Davoudi, S. and Healey, P. (1995), "City Challenge – a sustainable mechanism or temporary gesture?", in Hambleton, R. and Thomas, H. (eds), *Urban Policy Evaluation*, London: Paul Chapman Publishing.

Dowding, K. (1995), "Model or metaphor? A critical review of the policy network approach", *Political Studies*, 43, pp. 136–158.

Foster, J. (1997), "Challenging Perceptions: 'Community' and Neighborliness on a difficult-to-let estate", in Jewson, N. and MacGregor, S. (eds), *Transforming Cities*, London: Routledge.

Frances, J. et al, "Introduction", in Thompson, G. et al (ed.), *Markets, Hierarchies and Networks*, London: Sage.

Ghuman, P. A. S. (1980), "Bhattra Sikhs in Cardiff: family and kinship organization", *New Community*, 8(3), pp. 308–316.

Hahlo, K. (1998), *Communities, Networks and Ethnic Politics*, Aldershot: Ashgate.

Hall, S. (1996), *Race: the Floating Signifier*, Northampton, MA: Media Education Foundation.

Hannerz, U. (1980), *Exploring the City*, New York: Columbia University Press.

Hardill, I. And Raghuram, P. (1998), "Diasporic Connections: Case Studies of Asian Women in Business", *Area*, 30(3), pp. 255–261.

Healey, P. et al (1988), *Land Use Planning and the Mediation of Urban Change*, Cambridge: Cambridge University Press.

Herbst, S. (1994), *Politics at the Margin*, Cambridge: Cambridge University Press.

John, P. (1998), *Analyzing Public Policy*, London: Pinter.

Judd, D. and Parkison, M. (eds) (1990), *Leadership and urban regeneration: cities in North America and Europe*, London: Sage.

Khakee, A. et al (ed.) (1999), *Urban Renewal, Ethnicity and Social Exclusion in Europe*, Aldershot: Ashgate.

Levine, R. F. (2001), *Class, Networks and Identity*, Lanham, Maryland: Rowman and Littlefield.

Lorenz, Edward H. (1991), "Neither Friends nor Strangers: Informal Networks of Subcontracting in French industry", in Thompson, G. et al (ed.), *Markets, Hierarchies and Networks*, London: Sage.

Macpherson, Sir W. et al (1999), *The Stephen Lawrence Inquiry*, London: The Stationery Office.

McDowell, L. (1999), "City Life and Difference: Negotiating Diversity", in Allen, J. et al (ed.), *Unsettling Cities*, London: Routledge.

Moore, H. (1994), *A Passion for Difference*, Cambridge: Polity.

Nanton, P. (1998), "Community politics and the problems of partnership: ethnic minority participation in urban regeneration networks", in Saggar, S. (ed.), *Race and British Electoral Politics*, London: UCL Press.

Oliver, M. (1993), "Disability and dependence: a creation of industrial societies?", in Swain, J. et al (ed.), *Disabling Barriers – Enabling Environments*, London: Sage.

Pile, S. (1999), "The heterogeneity of cities", in Pile, S. et al (ed.), *Unruly Cities?*, London: Sage.

Powell, Walter, W. (1991), "Neither Market or Hierarchy: Network Forms of Organization", in Thompson, G. et al (ed.), *Markets, Hierarchies and Networks*, London: Sage.

Rhodes, R. A. W. (1997), *Understanding Governance*, Buckingham: Open University Press.

Rhodes, R. A. W. (1999), "Foreword: governance and networks", in Stoker, G. (ed.), *The New Management of British Local Governance*, London: Macmillan.

Riley, K. (1999), "Networks in Post-16 Education and Training", in Stoker, G. (ed.), *The New Management of British Local Governance*, London: Macmillan.

Skelcher, C. et al (1996), *Community Networks in Urban Regeneration*, Bristol: Policy Press.

Stoker, G. (ed.) (1999), *The New Management of British Local Governance*, London: Macmillan.

Thomas, H. (2000), *Race and Planning: the UK experience*, London: UCL Press.

Thomas, H. et al (1996), "Locality, Urban Governance and Contested Meanings of Place", *Area*, 28(2), pp. 186–198.

Thomas, H. and Imrie, R. (1989), "Urban redevelopment, compulsory purchase and the regeneration of local economies: the case of Cardiff docklands", Planning Practice and Research, 4(3), pp. 18–27.

Thomas, H. and Lo Piccolo, F. (2000), *Governance, Best Value and Race Equality*, Paper in Planning Research, Department of City and Regional Planning, Cardiff University, UK.

Thomas, J. and Ritzdorf, M. (eds) (1997), *Urban Planning and the African American Community. In the Shadows*, London: Sage.

Thompson, G. et al (ed.) (1991), *Markets, Hierarchies and Networks*, London: Sage.

Ui, S. (1991), "'Unlikely Heroes': the evolution of female leadership in a Cambodian ethnic enclave", in Burawoy, M. et al (ed.), Ethnography Unbound. Power and Resistance in the Modern Metropolis, Berkley: University of California Press.

Ward, S. (1994), *Planning and Urban Change*, London: Paul Chapman Publishing.

Werbner, P. (1991), "Taking and Giving: Working Women and Female Bonds in a Pakistani Immigrant Neighborhood", in Thompson, G. et al (ed.) (1991), *Markets, Hierarchies and Networks*, London: Sage.

Chapter 2

Minorities, Local Communities, Participation and Regeneration Processes in Inner Urban Areas

ELISE HENU

Participation and Ethnic Minorities: a French Paradox

In France, regeneration processes are often faced with a double problem: firstly, participation has always been a kind of political credo for distressed areas although urban planners have a lot of difficulty defining what it may really mean at ground level. Secondly, participation should be conceived and developed in areas where so-called ethnic communities live. But how do planners take into account this aspect in a country like France, where the words "ethnic" and "communities" are not traditionally used? After a short explanation of the French context and the presentation of my work on two case studies, the demolition of a shantytown and the rehabilitation/destruction of a social housing estate, I will focus on the relevance of such notions in the French context.

Setting the Scene: Politique de la Ville, French Paradox and Participation of Local Communities

A Few Words on the "Politique de la Ville"

The policies developed in distressed areas are supposed to take into account the resources of the inhabitants. In a traditionally centralized and representative political tradition, the difficulties of those areas and a wider context of decentralization have encouraged politicians and professionals to think about participation. The "Politique de la Ville", as an urban policy for distressed areas, more or less the equivalent of the British inner city policy, had to take into account this participative dimension, whilst changing its approach to public action towards crisis-stricken areas and towards the involvement of those areas in their own development.

At the end of the Seventies, the first processes of urban, economic and social deterioration of certain areas were noticed by various observers and public actors. It then seemed necessary to re-think public intervention, which was too centralized and fragmented by the administrative division of ministerial judictions, and to make it more transversal as well as closer to the ground. The "Politique de la Ville" thus became the privileged instrument of this new kind of public action, based on a logic of "projects" and "contracts", between central government and local authorities. The idea was, at the beginning of the Eighties, to spark off a partnership between the various ministries and their representatives, as well as a partnership with local authorities, within a renewed development philosophy.

Thus the "Politique de la Ville" was supposed to develop positive discrimination (a new, strange concept in France) and also ways for the local residents to express themselves and to become a part of the redevelopment process. Those participative policies had to be defined at ground level by taking into account its weaknesses and resources and the local residents' opinions about urban and social projects.

But, in a context characterized by the growing importance of local powers (the various decentralization Acts of the early Eighties thus acknowledged and developed mainly the mayors' power) and by representative democracy, it always seems to be difficult for politicians to share their power with the local residents or their voluntary sector representatives. Local residents' organizations are often limited to a consultative role as the idea of co-government is systematically rejected. The most interesting features of the "Politique de la Ville", which could be understood as community involvement, seem to be difficult to implement in the French context, especially if the communities involved are "ethnic" ones.

"Ethnic" Minorities and the French Paradox

In France, multiculturalism and community participation seem to be words and practices borrowed from foreign contexts. It is difficult for researchers and urban planners to address certain issues in terms of ethnicity since the assimilationist paradigm is really dominant. This does not mean that ethnicity has no meaning in social life in France! It is more like a kind of paradoxical taboo. Ethnic communities are not supposed to exist politically but they do in reality, being reduced to a sort of "negative" presence in political discourses about migrants and/or the integration of their second or third generations. Therefore ethnicity is often considered as a problem and not as an issue.

From the point of view of the assimilationist model, it is not supposed to be a part of social and urban planning policies. Indeed, the French political tradition considers nationality to be a decisive criterion, and thus erases all cultural specificity. The French political space was thus built on the ignorance, or even the destruction of local identities – which were particularly strong until the 19th century – as well as on an endemic suspicion towards any intermediary body based on this type of distinction. As for the successive migratory waves, they were received, politically speaking, in the same way as regional cultural specficities. Cultural differences, whatever they be, are rendered invisible by the dominant criterion of nationality/national citizenship. Thus to be French is to be equal to all others, regardless of origin, culture, religion, etc. These differences are not supposed to be expressed as such in the public space, or to become the object of specific policies, and are limited to the private sphere. Thousands of immigrants have thus been assimilated through diverse socializing institutions (e.g. work, school, army) and through French nationality. These people are supposed to be French citizens like all others, and are represented politically by exercising their right to vote, whatever their origin. There does not exist any community pre-established by its ethnic origin, at least according to the strict republican tradition.

Indeed, ethnic categorization seems to be functional in the social life of a country which regularly witnesses the re-occurrence of long debates about migrants, their children, their ability to assimilate, the importance of their origin. These categorizations do not seem to be limited to social representation, but appear rather to have a bearing on the practice of the various actors.

Such categorizations therefore attract the interest of an increasing number of researchers who study their various forms and consequences. They are quite an issue in themselves, in fact. Indeed, there is no statistical measurement including "ethnic" categories properly speaking. The only available data in this domain are those concerning the "cultural origin", a category which is only beginning to be integrated into our statistical methods, and one which causes much debate (perhaps with reason) because it seems to jeopardize republican equality. The emergence of such categories seems to endorse, indeed, certain representations and social practices, which are the sources of deep inequalities and segregation. Indeed, in the field of urban policy, in spite of theoretical republican equality, ethnicity seems to constitute an informal criterion for urban social policies. Yet many researchers in the field of urban policy identify certain trends of ethnic segregation due to informal influences on population by different kinds of actors, for example urban gatekeepers (local administrations, politics etc).

Thus, considerations about racial or cultural differences are not supposed to exist statistically and politically but they are part of everyday social, professional and political life. This is one aspect, for instance, of what we consider to be the "French paradox": a strong equalitarian national paradigm which apparently ignores all cultural, ethnic and/or racial distinctions, even though the latter are socially constructed.

About Participation: Ethnicity and the Crisis of Citizenship

As ethnic communities are not supposed to exist, they are not supposed to be involved as communities in the regeneration processes. Anyone with French nationality and voting rights is a citizen and is expected to use this power. They can also be involved individually in participative processes. Then they often have to face the same participation difficulties as the other tenants or local residents. But they also suffer from a situation of discrimination and segregation because of their ethnic origin and social-economic status.

Indeed, the emergence, in the late Seventies, of places signaled by processes of social exclusion sparked off a sort of crisis of local citizenship. The people who formulated the "Politique de la Ville" found that some people are supposed to be just normal citizens but certain complex urban and social mechanisms keep them in a situation of second-class citizenship. Participation then seemed to offer an efficient way of fighting against exclusion and of restoring a sense of "social cohesiveness". But how could "participation" be conceived when it seemed to be a typically middle-class attitude, and when the middle-classes had deserted certain areas of towns and cities, precisely during the Seventies and the Eighties? How can we conceive the participation of the populations in the most difficult situations, when they are precisely those hit the hardest by the crisis, whatever their origins? How, finally, can we talk of participation and citizenship when the most fundamental rights, like the right to work and the right to housing,[1] find no practical translation? These questions are significant of a general crisis of citizenship, which threatens the classic mode of integration of the last waves of immigration. The latter have become the subject of a recurrent political debate, as well as of certain practices, all of which tends to distinguish and isolate them even more. To the informal processes of segregation and discrimination mentioned above, one must add the difficulties inherent in participation in crisis areas.

How can we speak of community involvement in this context? Despite the difficulty of using the "community" ("communauté") within the French context, I do think that certain groups have got an interesting role to play in regeneration processes. But I also think that ethnic criteria are not always relevant to define those groups. As their observation in the context of urban regeneration schemes revealed, those groups appeared indeed to be defined by very different criteria from those of origin or common culture. The resources mobilized by those groups, and their transaction capacity seemed to depend, among other factors, on their relationship with a given professional and political context proper to the locality observed. I found that the most important aspect in participation seems to be the quality of the relationship between the local community and political powers.

Two Case Studies and So Many Questions...

By working on two case studies in Marseilles, I have been focusing on different kinds of questions. I have chosen to work at ground level, using methods of participative research. By working extensively on those case studies (three months for the shanty town and more than one year for the housing estate), I was able to build personal networks with the local residents and share some day to day life with them (social and education work with the children, for instance). Through this "living" research, I had to reconsider my first theoretical positions and the ways I was addressing certain problems. These changes were due to the complexity of these case studies. Regeneration processes and "ethnic" minorities concerned both localities. They were both situated in the northern distressed areas of Marseilles. Local "ethnic" residents had suffered from the same deprivation processes than the other residents but their origins seemed to be another handicap in a difficult context.

Marseilles is the second largest French city. It is situated on the South East Coast. It was an important colonial port that has suffered from an economic and social crisis, deprivation processes and demographic decline of decolonizations (during the Sixties) and the crisis of the Seventies. Some areas have become typical pockets of deprivation. Rates of unemployment have increased up to 50 percent in certain social housing sectors. "Foreigners", or more exactly people with foreign origins, started to become concentrated in some parts of the city, which would be quickly but wrongly considered as ethnic ghettos. Indeed, in those areas deprivation processes were due to income-selective processes more than ethnically selective processes, even though the latter do exist. In more of the distressed places, even if many searchers identify trends of ethnic segregation, we have to use the word "ghetto" very carefully since the ethnic composition of the population is quite varied. In "ethnic" terms, people from all origins live in these areas: "ethnic" French as well as people of "migrant" origins, from other European countries or from Black Africa, North Africa, the Middle East, Asia (the last waves of immigration, from the Sixties). The common feature in these situations does not seem to be the foreign origin but the economic and social difficulties.

From the Shantytown...

Given these general considerations, the first case study could be considered as an exception. The shantytown that has been studied was named "Lorette" and was inhabited by a population of Algerian origin, more precisely Kabyle (from an eastern region of Algeria). It was pulled down and a large shopping center was built on the same location, an operation that was viewed by politicians and urban planners as a means of regenerating the area. While working on the first case study, the primary goal was to understand how people would react to an urban planning operation, namely, the destruction and relocation of their homes. For me, the research began by trying to understand the backgrounds of the residents.

I found that the residents all came from the same village in Kabyle ("Bouckhélifa"). They arrived in France in the Fifties, at first without their families. In a context of housing shortage, their employers (Marseilles Tileries) had not provided them with housing, only small pieces of land near their factory and its clay quarry. They were forced to build houses for themselves and their families to live in. Indeed the latter had joined the workers during the early Sixties. Therefore, this place had been occupied for thirty years by this Algerian community and was something like a Kabylian village in the heart of Marseilles. In spite of the economic, urban and social transformations, closure of factories, loss of jobs, etc., during the Eighties, these families have resisted social exclusion and persecution thanks to community solidarity. This exclusion within the community has resulted in the complete isolation of the neighborhood, which in turn, has become an ethnically segregated sector of the city.

Since the very beginning of the Eighties, when the first "Politique de la Ville" programs were put in place in the northern areas of Marseilles ("contracts" between central government and the municipality to finance urban regeneration and local development schemes), the shantytown of Lorette had been targeted for "renewal" in several projects. Those were not carried into effect, probably for lack of strong political determination and because of the focus on other places of this area, which were deemed to be more problematic (the "Plan d'Aou" estate had already been included in the renovation programs then scheduled). A phase of expectation then started for the shantytown, between rumors of demolition, which were never followed by action, and the demands of the children born in France and now adults to access to a certain degree of comfort. Only in the late Eighties did things get a little clearer: the projected building of a shopping center, "Nord Littoral", was to set off a dynamic of development in these deprived districts. That entailed a number of planning and development operations, among which the removal of the shantytown, situated on one of the planned accesses to the shopping center, and the building of a social housing estate in order to re-house the population of the shantytown. The re-housing operation submitted ninety families to a process of urban renewal and social work. The people of Lorette were relocated in individual houses, in the new social housing estate, in 1995–96. It was really a different life from the shantytown and they had to live with the new status of tenants, which they had never had.

Starting with the above knowledge, the research work began in 1996 with the study of the effects of this urban planning operation on the lives of the residents. However, after meeting with the families and discussing re-housing issues, I discovered that in addition to the radical change in their day to day lives, the inhabitants were equally concerned with the negotiations which were being held in order to complete the project; negotiations which determined how, where and when the residents would be displaced. The question of the empowerment of local residents thus became an obvious matter of concern, although the group of people was definitely at the lower end of the social ladder. But they did not like to be planned for, they wanted to take part in the process. This attitude was above all significant of the inhabitants' concern for their own future, in an urban development operation in which the stakes – as they were well aware – went far above and beyond the destruction of the shantytown. They were worried about the way in which re-housing operations would be carried out, and about the way they would cope with their new status of tenants. Their concern also went to the survival of the communal entity of which the shantytown had been the spatial and physical basis. Indeed these people had been living in a Kabyle "microcosm" since the Sixties, a microcosm which had reproduced a number of spatial and social patterns imported from their Kabyle origins. The actors involved in the process of urban renewal were thus confronted with the expression of all sorts of demands and expectations, concerning for example the level of rents and expenses, the management of new public amenities, the level of compensation for the lost houses of the shantytown. Moreover, with these same demands, essential questions were also formulated, such as the future of the community. Very soon, the residents of Lorette expressed the wish to "stay together", and not to be scattered in the neighboring social housing estates. The problem then was to maintain community networks on which daily life – or even survival – was based, as much as to avoid the more obviously depreciated forms of social housing. The neighboring social housing estates (including "Plan d'Aou") had indeed a very bad reputation, and were considered by the people of the shantytown as "damned" places, where the children would "rot", and from which the families and the community as a whole should be protected.

The professional and political actors of the clearance program were thus very rapidly confronted with these demands. There was not only the Marseilles council and the mediation team which it had put in place on the ground (M.O.U.S., or "Maîtrise d'œuvre Urbaine et Sociale"), but also other public operators like the social housing contractor, "Habitat Marseille Provence" (H.M.P., the main social housing contractor in Marseilles). The developer of the shopping center, on the other hand, was a private operator in land development by the name of TREMA. Political and institutional actors were therefore particularly fragmented, and this diversity was to be at the root of different attitudes towards the people of Lorette. Those closer to the ground were well aware of the residents' problems, difficulties and resources, and realized that their demands were well founded. They therefore encouraged a whole process of mediation and negotiation through which most of the expressed wishes were fulfilled. However, pressure on land development, linked to the very nature of the operation, was to make these attempts difficult to sustain over time. The clearance of the shantytown, and the re-housing of its population were carried out within five years only, which represents a rather short period of time for a complex operation. Certain actors, because of their remoteness from ground operations, or because of their own interests (notably those of the developer), were little aware of the amount of time necessary for such mediation processes to take place. The pressure increased on the local residents and actors on the ground, to the point that the end of this operation was marked by an intervention by the police to evict some residents who were not satisfied with the amounts offered for compensation and therefore still occupying the site of the shantytown. These few authoritarian re-housing operations "spoiled", so to speak, the careful mediation and negotiation work, which had accompanied the whole development process. They also revealed something about the way in which this urban renewal operation had been defined and roles shared between the different actors, and about the recognition of the residents as potential actors among all others. At a certain moment in the process, the stakes of the residents themselves seemed to become out-weighed by the stakes of the operation proper, and the community's logic was then overtaken by the logic of development all the more so as the residents were not considered as relevant partners with urban planning needs anymore.

As a result of these findings, I started to become interested in the topic of urban planning negotiations during regeneration processes: negotiation being in this case defined as a sort of ongoing transaction between actors with specific resources and scope for action which are constantly redefined all along a given process. The case study of Lorette had led me to understand that, even – and perhaps especially – in a really distressed situation, people had some interesting resources that enabled them to become a part of those processes even if this was not fully recognized by political or professional authorities. When I started to work on the second case study, the operation of destruction/rehabilitation of a social housing estate, I realized that this ability to negotiate was also used in a quite different context.

... to the Social Housing Estate: A Similar Question, the Negotiation of Urban Developments by Local Residents

This second case study was quite different from the first one. The social housing estate of Plan d'Aou was situated in the neighborhood of the shantytown, near the new housing estate, which has been built to relocate people of Lorette. The estate of Plan d'Aou had been built at the beginning of the Seventies, which were a period of intensive building of social housing. It was composed of 925 flats and was fully occupied up to the beginning of redevelopment in 1986.

Thus it was quite a different context from Lorette. One of the first elements is of course the difference in scale between the two operations. Obviously the clearance of the Lorette shantytown concerned "only" 90 families, when the first operations in the Plan d'Aou estate had to deal with ten times as many people! Beside this question of scale (in terms of population) is another important difference, that of the status of the residents. The residents of Lorette were considered, according to French law, as "legalized squatters".[2] The residents of Plan d'Aou are tenants. They rent their houses from three different social housing landlords who own the estate between them. Also, the residents of Plan d'Aou had been affected by Politique de la Ville measures much sooner than those of Lorette, because of their status as tenants, and they had already met the whole range of actors (from social worker to urban planner) which this policy mobilizes on operation sites.

Finally, the tenants did not have the same community history: the population had changed during the crisis of the Seventies. Indeed the first tenants of the estate belonged to the middle and working classes. But the population of the estate had become poorer and poorer as the wealthier tenants were leaving the estate and the poorer new comers were renting the newly vacant flats. This process can be explained as much by the new opportunities offered to the middle classes, notably in terms of property ownership, as by the opening up of social housing estates to much better-off social categories. The new tenants of the Plan d'Aou estate came, indeed, as soon as the late Seventies, from cleared peripheral shantytowns in North Marseilles or from rehabilitated areas of the center. The people coming to the estate were thus not only the less better off but also the most pauperized of this urban region. As more or less consenting "victims" of the land market pressures in other areas of the city, many of them were discovering a form of social housing, which was still a symbol of social improvement. Amongst these very poor people, the percentage of those who arrived in France with the Fifties-Sixties immigration was particularly significant, and was even to increase with new income-migrants, whose origins varied according to the economic and political context (Turkey, Vietnam, the Comoros). Like many social housing estates, Plan d'Aou started to become characterized by high rates of populations with foreign origins such as Comoros, Algeria, Tunisia, Morocco, Vietnam, Armenia, Turkey (rajouter composition par taux à côté de chaque pays). Today, it is still a kind of multi-ethnic village that is currently being partly demolished.

The differences between Plan d'Aou and Lorette are thus numerous, and are due as much to the histories of both areas as to their scales and social composition. They are, however, two neighboring areas geographically and are therefore faced with the same urban pressures (notably the development of the retail center, causing an increasing land-market pressure) as well as with the need to negotiate. Therefore, the local residents, despite their different histories and origins, want to be a part of the process to make sure that their interests are taken into account.

Since the first rehabilitation strategies, in 1986, urban planners have started to reduce the population density of this estate. In 1985, 925 families were living there. The estate had a really bad reputation because of problems of violence and drugs. It had been completely abandoned by the three social owners, and it was like an urban village with its own rules and solidarity based on an informal boundary between the inside and the outside. A sort of law of silence and honor, something like an "Omerta", seemed to structure the social life of the estate. It was in this particular context that the political authorities started to work on the renewal of the area. One of their first purposes was to reduce the number of tenants and to destroy the less accessible part of the estate. Their second and informal purpose was to isolate the most famous and "great" families that were using illegal ways (drugs, stolen cars…) to survive. We can say that the reasons for urban planning in this context were less physical and/or environmental planning than they were in the case of Lorette.

Perhaps these reasons could be owing to preoccupations inherent in "social peace", with all the approximations which such a notion may convey in the assessment of an estate's situation. Moreover, it is very likely that those law-and-order preoccupations have served the various interests in the land and housing markets, evident in this area by the Eighties. When the retail center was only in its draft phase, certain large estates were already being considered for renovation or even destruction. In the case of Plan d'Aou, the renovation process seems to have followed a timid evolution, which can be divided into several periods.

From 1986 to 1988, the Plan d'Aou estate witnessed the destruction of the blocks farthest away from the estate's main entrance. The space locally became less "dense". Tenants were generally re-housed in vacant flats in other tower blocks. But these first destructions were not followed by construction operations, nor were the remains of the demolition site cleared away. Until 1993, the residents of Plan d'Aou lived among the ruins of the blocks, which had been pulled down. The process appeared to have come to a halt, and "normal" progress in terms of urban policy was only restored by the determination of the council to build the retail center. The years around 1993 saw new surveys carried out, aiming to better identify housing needs and above all the population present at the time. The latter had significantly decreased. Indeed, there were now only 400 families left. The out-migration could be explained both by the departure of people whose flats had been pulled down, and who did not wish to be re-housed within the estate, and a deteriorating climate. After so much time, many tenants did not wait until the planning and development projects came to fruition. Their departure served the interests of the social landlords, who were implementing a policy of vacancy: the abandoned flats were often walled up, to avoid squatting. Moreover, the building of the neighboring retail center, a prestigious project, led the remaining local residents to think they were becoming undesirable, because of their social condition and various origins. The project as it was then did not prove them wrong: the possibility of pulling down the whole estate had been considered, and re-housing proposals were rather hazy as to where new flats might be located. All these elements put together smacked of planned gentrification, in the longer run. "They're trying to push us out", the inhabitants said, while the council's heavy and ambiguous silence did nothing to reassure them. The council's plans lacked clarity indeed, and they seemed particularly uncertain and subject to constant revision. As for the social landlords, they appeared to favor a strategy of land value management, which indeed consisted in encouraging the destruction of the estate through the development of vacancy. Within this context, the reaction of the inhabitants followed both the way of formal negotiation (questions were put to the council by local councilors, calls were made to the government...) and that of more informal bargaining (such as attending council meetings, occupation of social landlords' and the council's offices, various pressures, verbal violence, at times, against social workers). In the end, the 1993 surveys were not to produce further effects. They were resumed in 1998, and the program reached its operational phase only in November 1999.

The program still mobilizes the same public actors: the municipality, the three social landlords and the government as financier. The new project consists essentially of the destruction of the highest towers, the rehabilitation of the smallest ones, and the building of new social housing units for the priority re-housing of the estate's residents. The latter measure is one of the consequences of the residents' negotiations. Residents of Plan d'Aou, beyond their concern for the social – and perhaps informal – aspects of the public actors policy about their estate, wished to maintain a number of community networks which, as was the case with Lorette, help them to survive on a daily basis. Other claims and complaints also marked this process: especially about expenses, the management of public spaces, for offices and subsidies for associations, for devices against dumping inside the buildings, extermination of rats.

Thus, after the destruction of five blocks of flats and the construction of the first 90 new units, the renewal operation now concerns 275 families who still live in their original blocks. Those families have been living in Plan d'Aou for years and for most of them since 1975. They know the history of their estate and they have suffered for years from bad living conditions including the context of insecurity, the uncertainty regarding the renovation projects, and the poor quality of the buildings. Despite the multiplicity of origins, whether urban or migratory, the inhabitants of Plan d'Aou thus share, and have been sharing for years now, the same founding myths about the estate. They are "from Plan d'Aou", before being from Marseilles. An identity was thus defined in a more or less positive way, through opposition to the outside world and/or through a shared stigmatized history. This identity encourages their endemic – and probably justified – distrust of the political and administrative bodies in charge of rehabilitating the estate. The residents often ask the said bodies about, for example, the reasons and the means for rehabilitation. Among other topics, the plans for dispersing the locals are always mentioned. The question of gentrification seems to be more present than ever. Indeed, the newly built units, of better quality are humorously referred to as "Beverly Hills". This refers to their well-finished aspect but also connotes the suspicion, ever present, that these flats might be destined to others, even if they are currently rented by families from Plan d'Aou. The estate community thus produces a whole argumentation *vis-à-vis* the actors of urban renewal, which includes both its own identity and elements of a class and/or territorial struggle.

Negotiation, Communities, Ethnicity: the Multiplicity of Forms

Thus, regarding the question of negotiations, we found once more that the residents had the same kind of concern. They were worrying about the re-housing conditions and the way they might be disadvantaged by the urban operations. Those negotiations have allowed each and every actor to defend their stakes in a regeneration process. The residents' stakes could be defined as the way they could be re-housed without losing their day to day life and above all the community networks that help them to survive. The re-housing operations provide many alternatives that are precious for local residents because they define conditions of a new way of life in a new neighborhood. In both case studies, we found the same kind of mobilization processes, with the structuring of formal as well as informal groups taking part in the negotiation processes in different ways. This structuration generally predates the urban renewal programs, but finds in them an opportunity of becoming more formalized and reach a higher level of public debate.

In the case of Lorette, residents' mobilization existed in the early Eighties, when the redevelopment project was only in its first phase. Before then there existed a youth organization with social and cultural purposes. Its ambition was to offer a number of leisure, social and cultural activities to the youths of the shantytown. It was on the basis of this existing association, which didn't have official status, which the young people of Lorette began to speak out about their future and that of their parents. The young expressing themselves was a reflection of their increasing power within the shantytown, at a time when their parents could not take part in the process, for lack of command of the vocabulary and of the rules of the participative game. Many of them were not at ease with the French language, and adopted a "respectful" and fatalist attitude towards the institutions. In a certain way they were reproducing the employment and colonial pattern of relationship which had dominated their lives, in France as well as in Algeria. Their sons, not having known the colonial period, and having always lived in France, did not adopt the same attitude at all towards the actors of urban renewal. In the case of Plan d'Aou, various associations also existed before the rehabilitation programs, on the basis of common interests (with, here again, a generation divide), or common origins. This association microcosm gave birth to the "Tenants' Friendly Society", which was dominated in this case by the middle-aged men of the estate, who were more equipped for such dealing than their shantytown counterparts because they were more "socialized" through their voluntary sector experience. Their relationships with the younger residents' of the estate often take on the form of a conflict of generations, which was not absent from the world of the shantytown, but was not expressed as such in terms of associations. Thus, there is now in the estate a "Youth Association of Plan d'Aou" intent on making its voice heard in the urban renewal process. However, it only has a very limited audience, for rather complex reasons, which will be explained later.

As has already been mentioned, negotiations in the case of Lorette were primarily oriented toward re-housing conditions, the guarantees that could be given to the families in particular, and finally the preservation of the community itself. The more active members of the group had managed to become represented in the political bodies concerned, at the highest local level, as well as in the sphere of field practitioners and professionals, through whom a number of their demands also made their way up. The rather exceptional character of the group as well as a form of "sympathy" on the part of the professionals involved allowed the said group to access the political arena on which its future depended. In much the same way, the political support for the project, as well as the intensive mediation work carried out brought families and decision-makers closer together. If this pattern of negotiation, which might seem ideal, eventually collapsed, it was under the pressure of land development interests, which accelerated the process by imposing the re-housing of families. Well aware of this state of things, the young of Lorette then adopted more radical tactics and moved to more virulent forms of action: occupation of the building site of the new social housing estate, demonstration and blockage of the traffic around the retail center under construction, refusal of re-housing proposals and occupation of the shantytown houses... until their eviction by police force. The situation became very tense and resulted in a somewhat authoritarian style of planning, as attempts at mediation failed because of the differences between the professional actors, and, indeed, their divisions. The clearance of the shantytown of Lorette and the negotiations which accompanied and followed it, appear like the resistance of an entire village which might have won over the favors of certain actors whose interest it was that things went along as smoothly as possible. The project of the retail center was an important one indeed, and the majority of the actors involved could only have benefited by the involvement of such a community. The conditions of the negotiation started to deteriorate as soon as the consensus about the status of the negotiation itself began to crumble and lose its strong political and professional support. The existing relationship with the Lorette residents lost in quality what it lost in recognition and reciprocal "readability". During this last stage, operational workers and professionals in the field, who had also supported and "carried" the negotiation process, found themselves becoming gradually less recognized than other actors, their legitimacy and negotiation potential being therefore questioned.

The story of the negotiation in Plan d'Aou is a little more complex. Differences in scales and history of the site have significance here, as well as the political support for the project, and lastly the representation of the local population by the actors of urban renewal. The case of Plan d'Aou is in fact much less exceptional than that of Lorette. Plan d'Aou is a social housing estate, which has undergone several "social development" programs since the early Eighties, being included in the various "Politique de la Ville", designated areas. The local residents are used to contacts with the administrative and political spheres because their daily life has been marked, from the onset, by dealings with the public sphere, notably through the "social landlords". The rehabilitation program therefore is in no respect really exceptional, all the more so as it is scheduled over a long period in terms of decision-making and implementation. It is one of the elements through which the relationship between the estate and the surrounding political and institutional environment has known a certain continuity. The residents' claims bear witness indeed of a long-established relationship, of the importance given to questions relating to the management of collective, daily-life space, but testify also of the residents' knowledge of professional and political actors and their logics. The suspicions about the unofficial "social development" program concerning the area arose because of previous relationships and knowledge from which the shantytown could not benefit. In other terms, the clearance of the shantytown concerned a locality, which was autonomous until then, and was organized within the framework of an urban renewal program, by actors with whom the shantytown's residents had never dealt. The rehabilitation of Plan d'Aou may rather be located within the history of the relations between institutions and inhabitants within a space, which was by nature heteronomous. Not surprisingly then, the way in which the negotiation was structured was partly influenced by this public history. During the first phase of rehabilitation operations in 1986–1988 the associations present in the estate (this time in a formal manner: those associations were registered in the social, sports and cultural fields, etc.) had been fully integrated and taken on board. The tenants had nevertheless preferred to found a "Plan d'Aou Tenant's Friendly Society" also formally recognized, and affiliated to a powerful national federation of tenant's associations, the C.N.L. (National Confederation of Tenants) in order to defend their interests in the housing and urban planning sectors, with the use of legal aid and of various networks of political influence. This attitude shows they have a better knowledge of their political and administrative interlocutors and of certain informal levers. This Tenants' Friendly Society has had a crucial role in the negotiation of rehabilitation operations. But it rests on a totally different basis from that of the Lorette association, using other forms of influence and self-organization, in direct relation to the way that this estate and its rehabilitation have been conceived, from a political as well as a professional point of view.

In the case of Plan d'Aou, we saw that the institutional context and the objectives of the operation are very different from those of Lorette: long-term rehabilitation/ renewal projects, not always followed with effect, and what is more, a climate of suspicion as to the real motivations behind those projects, a "social peace" imperative. Those numerous elements of indetermination leave the local residents with many uncertainties concerning their future in the estate, but it is also true that this situation affords them a certain amount of power and potential for action in the negotiation process. Being aware of the bad reputation of their locality, they are suffering from it but they also know how to use it as a way of putting local authorities under pressure. The fear of urban riots (which have already happened in large French cities but not in Marseilles) is a constant feature of the approach to those areas of many politicians and professionals. The threat of urban riots is an argument, which plays an important role in the negotiation process, which was less powerful in the case of Lorette. In this context, we are dealing with a new form of power, which could be defined as "mass" power. This mass power knows perfectly how to use any political indetermination, and above all the numerous forms of mediated representation to which the actors of urban policy may give occasion. In this sense, the institutional vision of the inhabitants of the estate, a vision which often emphasized themes related to insecurity, is used and instrumentalist within the negotiation process, as arguments and means of pressure, by the various formal or informal[3] groups themselves.

The inhabitants' awareness of the representations and the fears they sustain, especially for those of the actors not working on the ground, is limited to a long past of cohabitation. The relationships with the professionals involved in the life of the estate were built little by little through time. Their management of the estate has often been put into question, their plans interrogated rather crudely, especially since the start of the rehabilitation program. The circulation of information within the estate, often through informal channels and interpersonal networks, served nevertheless as the basis for a discussion of the planning projects, with different interests emerging, following the various formal or informal groups concerned. This history determines therefore the relationships between the residents of the estate themselves and their modes of mobilization, formal and informal. Something began to take shape little by little, which did not exist in the case of Lorette, by virtue of the somewhat exceptional character of the residents' mobilization: the formation of a sort of local "middle class" in the estate, with relationships with the various actors on the ground, and retaining thereby a certain amount of power over the other residents, even if only symbolically. This stratum sees a form of social upgrading in any sort of negotiation, by having access to networks that open up new possibilities. It includes all the people in charge of associations who have reached a level of power significant enough within these to become the privileged partners/

interlocutors of the political and professional actors. Equally included in this "middle class" are the leaders of informal groups who manage, with the help of the formal groups, to make themselves heard by the institutions. Finally, also present are local people who have found a place in certain urban and social institutions in charge of the management of the estate's daily life (social/cultural "animators", youth social workers, building caretakers...). All these actors, constituting the estate's "middle class", share together the values conveyed by the professionals, the politicians and the technical or engineering staff who deal with the estate. The last groups can be seen as real institutions of socialization, with values, a language, and priorities, which reflect their own professional sphere as well as their close contact with the estate. A whole fringe of local residents is thus constituted, the "thinking class" of the estate, so to speak, by people who recognize and assert themselves as such, and find in the parts they play with the professionals many occasions of getting recognition for this "special status". In this complex interplay of relationships, the professional and political fieldworkers[4] play a significant part. They allow a whole sphere of proximity to take place, and are the instruments of communication between the political, the professional and the local residents' spheres. But this universe seems to come into existence only after a long period of relationships, as in the case of Plan d'Aou. And, what is more, its appearance brings with it no guarantee of a better participation in the negotiation process.

Thus, when taking stock of the existence of such groups, one can only recognize the even greater complexity of this social housing estate, along with the increased difficulty of reading the negotiation processes. The expression of the various claims and grievances is indeed torn between the "exact", "immediate" representation of the residents and various interest groups sharing the "privilege" of relating to the institutions. There is not, therefore, one but several forums within the estate, which could be described as a series of concentric circles centered around the access to the professional world of planning or social work. The relationships between the residents and the institutions, and the very object of the negotiation, become much less clear, and the negotiation process itself particularly complex to analyze. The Tenants' Friendly Society, for example, is one of the expressions of the local "middle class", which is in fact complex, structured by groups of varying influence (according to age among other factors) conveying its own claims and whose representativeness is not always certain. It enjoys, however, certain recognition, due to the local history, by the social landlords and the political actors, even if that recognition may at times be limited. But the expression of other groups (the young, for example) who have not yet had access to this political forum, is much less taken into account, and depends on the good will of the representatives of the "middle class" towards them. The "middle class" seems to act as a kind of buffer which, following the quality of its relationship with professional or political spheres, may delay the negotiation process, distort the information from both sides, while purporting to represent local residents.

Moreover, the porosity between the professional sphere and that of the local residents is not necessarily a guarantee of real communication. The "middle class" does not always have as much attention as it would like from the political and institutional actors. Compared to the other residents, it seems favored by its proximity and privileged links with the institutional sphere, but it remains nevertheless a weak partner in a negotiation process, which may include the local residents only if there exists a certain political support. The Plan d'Aou operation unfortunately suffered, like many others, and contrary to the case of Lorette, from a political confusion as to its objectives. It has also been marked by the often negative representations of the inhabitants of the estate, representations, which never encourage power-sharing. The "mass" argument and the ever present threat of riots and urban violence are sometimes worth considering, especially in states of emergency; they remain nevertheless discouraging arguments with regard to participation in the long run.

Thus, the few elements of indetermination left to the residents in fact turned "against" them, by never defining their potential for action in this process. The negotiation then evolves like a series of desperate attempts to gain a little more attention. The demonstrations of verbal violence, or, more rarely, physical violence, or the lock-out of official buildings are acts of protest addressed to the actors of urban renewal, with a view to reminding them of potential threats, and gambling on an "echo effect". These protests are also more often than not undertaken by persons or groups who disagree with the more customary and recognized members of the "middle class". Thus, where the possibility of expression is not given, or where attention to such expression is not paid, the situation becomes very conflictual. The management of such conflict then become a matter of urgency.

In both case studies there is assuredly community mobilization of community resources. But the forms these take are radically different following the history of the localities and the quality of their links with the political sphere. For Lorette and Plan d'Aou, it would seem that the potential accorded to community resources and mobilization largely depended on the political and institutional context of these programs. One common question remains: who are the groups of actors? What is their nature?

By taking into account this striking similarity, and the elements concerning the negotiation processes in themselves, I had to go back to another aspect and to reconsider an equation used during the study of Lorette, perhaps too easily. I had thought that negotiations were "ethnic" because they were dealing with identity components and mainly the wish not to be dispersed. It was easy to think because there was one place, one ethnic community based on the same origin, the same history, the same migrant trajectory: ethnic community = identity = negotiations. I realized that the same arguments could be used in the case of Plan d'Aou in spite of a multi-ethnic situation. Our questions about negotiations had to be further adapted: What is the role of ethnicity in negotiations? Do cultural identity, common origin or nationality (criteria usually supposed to define ethnicity) result in identity factors in negotiations? And, what kind of negotiations do urban planners have to face when they have to implement social urban policies in supposedly "ethnic" localities? What does it mean to have an ethnic bias in urban planning negotiations? *In other words, to what extent do these two case studies enable us to illustrate the French paradox? And what are the implications for negotiation processes in urban planning/development operations?*

A Few Conclusions...

A few points could sum up what we are able to conclude after working on these case studies:

Community Resources do exist and they can be an Interesting Alternative for Urban Policies

Working at ground level, we have discovered that communities involved in urban planning were characterized by interesting capacities for adaptation, mobilization and negotiation. Although people have to adapt to a new neighborhood and new living conditions, they can use the solidarity mechanisms developed during the constitutive phases of the community. The community can be considered as the group of people in itself but also as the history it is supposed to share in common. In distressed areas, this common history can become a real resource to face the difficulties of everyday life. We could find this kind of structure in Lorette but also in Plan d'Aou. In the case of Lorette, for instance, people were able to adapt to a new urban form and to a new status of tenants because they could keep their former way of life and adapt it to a new urban form. For instance, even if certain relationships were destroyed by the negotiation processes (rivalry between families) and even if certain community events were put in question by the new way of life (family and community celebrations were difficult to adapt in social housing, for instance), people from Lorette had kept an important group identity. This identity allowed them to adapt to a new neighborhood by still asserting themselves as people from Lorette. There was a similar phenomenon of group awareness in the case of Plan d'Aou. In both case studies it has enabled local residents to fight in order to resist dispersion in the city or in different social housing estates. That was an important aspect of the negotiations. People from Lorette and Plan d'Aou do know what their needs are and also what the stakes of the urban operations are for all the other parties involved. When facing those stakes, they perfectly know that the group is probably the most efficient level of negotiation. In the case of Plan d'Aou, the group is structured by various parameters, such as the relation to institution, origins, and common interest... but there remains the essential capacity to mobilize a community as a whole. This emphasizes the French paradox mentioned in the first part of this chapter: communities exist and they are important forces in urban processes although they are sometimes not very well recognized as such in the political game.

From Fear of Conflict to Perverse Effects?

Political and administrative systems are not supposed to consider communities as these systems do not recognize conflict as a working method. Therefore, although participation is supposed to be encouraged in distressed areas, one of its possible and sometimes main aspects, namely conflict, is rejected or badly managed. Participation then becomes a matter of emergency: responses are given to little-known localities through measure taken out of principle towards the closest and loudest groups. This does not mean that these groups are allowed to become more permanent parts within a participative device. This attitude may lead to some perverse effects in the social life of those localities. The professional and political practices encourage informal power struggles during the regeneration processes. The power of groups is informally involved in public decisions and those practices lead to a strange situation. People who might emerge as community leaders for regeneration processes may earn a little legitimacy in their negotiations with professional and public authorities. But, considering the difficulties they face to be totally recognized by urban planners and politicians as relevant partners, they can not use this legitimacy for social changes and their own social ascension. Public resources, which are spread financially or symbolically on those localities, become stakes of power between residents without any possibility of formal social recognition from outside the community. The case of Lorette, at the beginning at least, shows that the clear and strong political support of a group and its leaders not only enables the latter to become valid and valuable participants, but also allows for a clarification of negotiation rules. What is at stake then is the quality of the link between all the actors responsible for urban renewal programs on the one hand, and the ground level on the other. How are the different forces present discerned? How are they evaluated? And how are they taken into account in the negotiation process?

More precisely, what about "ethnic" communities in this context? We concluded that the notion of ethnicity is really an interesting object of study provided we can use it to focus on the way that the various actors consider such groups as culturally different, rather than focusing on the supposed cultural difference in itself!

Thus communities can exist and create their own identities in spite of ethnic differences within one locality. In the case of Plan d'Aou, there is no one single culture or origin. But a community can exist thanks to common interests, and/or a common history in a common place. This can be the reason for mobilization, and this mobilization can become a part of the group identity. It seems that identity should not be considered as a state but as a result of an on-going process. This process always evolves with day to day events and especially during periods like urban regeneration operations, which enable the group to establish a difference between "Them" and "Us".

Even if a common origin or culture exists, acculturation processes erase the distinctive cultural aspects. For instance, Lorette people (and above all the second and third generations) did not have the same cultural references as their parents. They were French in their nationality and in their way of life and a large part of their identity. And this factor precisely enabled them to be socialized enough to partake in the negotiation process. This socialization makes it possible to master enough common rules to take part in a dialogue. These common rules are numerous: the easy command of a language, the application of the most elementary interaction rules, a certain competence through the knowledge of the formal and informal workings of a local society. Without these competences, no encounter is possible. No ethnicity either. Indeed, one of the main aspects of those issues is the existence of the group by itself and above all the way this group is considered by different kinds of actors. Many professionals and politicians have a certain preconceived idea about the supposed "ethnicity" of others. It seems that ethnicity is defined not only by cultural heritage, but also by one's interpretations of customs and beliefs. The word "ethnicity" encompasses both cultural and social influences and is the result of one's interpretations of such influences. These interpretations are very valuable in order to understand the way professional and political actors take into account the facts and the people, which form part of the urban regeneration process. This last aspect can be linked directly to the themes of ethnicization and racialization, which at the end of the day are not related to an ethnic "state" but a whole set of representations centered around a supposed ethnic or cultural difference.

In view of all the above remarks, are ethnic criteria really relevant to study the participation and regeneration processes? Clearly, it depends on the definition one chooses of these ethnic criteria. They seem difficult to determine according to an overly "static" conception of ethnicity. It appears appropriate to consider ethnicity as a process, encompassing both the observation and the awareness of the observer of a supposed and/or acquired identity. In this way ethnicity can be seen to constitute one of the elements enabling groups to evaluate and represent themselves, and further to define and utilize their respective potential for action. Ethnicity therefore constitutes one of the elements, which are both essential and secondary in negotiation processes.

Participation seems therefore to be possible as long as a group exists on a basis that can be a common origin as well as a long common history. Participation is also possible as long as this group is recognized as a relevant group for public and political actions. Consequently, the main question seems to be that of citizenship and the recognition of formal or informal groups in planning processes, whatever their basis. But the recognition of communities is dependent on wider political considerations about participation and issues of ethnicity. These aspects seem to influence, following very different modalities, the quality of the relationship between the local community and political and administrative powers. The quality of this relationship seems to be a decisive factor when working with any system of negotiation, and in particular to fulfil the empowerment and citizenship ambitions of programs such as the "Politique de la Ville".

Notes

1.   As written in the Declaration of Human Rights, the preamble to the Constitution of the Vth French Republic.
2.   "Occupants sans droits ni titres", "Occupants without rights nor titles".
3.   By formal groups, we mean all legal associations, i.e. officially registered and publicly recognized as associations. By informal groups, we mean all groups of people, more or less stable, associated according to their interests, aspirations, etc., but without any form of official recognition and therefore not endowed with legal personality.
4.   We consider as fieldworkers all those whose occupations, or functions, put them in interaction with the residents, and whose knowledge of the "grassroots" is linked to their professional or political practices, but also to a whole network of interpersonal relationships.

References

Balibar, E. and Wallerstein, I. (1988), *Race, nation, classe. Les identités ambiguë*, Paris: La Découverte.

Barth, F. (ed.) (1969), Ethnic groups and boundaries: the social organization of culture difference, Bergen, Oslo: Universiteforlaget and London: George Allen and Unwin.

Bastenier, A. and Dassetto, F. (1993), Immigration et espace public: la controverse de l'intégration, C.I.E.M.I., Paris: L'Harmattan.

Beaud, S. and Noiriel, G. (1990), *Penser l'intégration des immigrés*, Paris: Hommes et Migration, 1133, pp. 43–53.

Blanc, M. (1998), "Social integration and exclusion in France: some introductory remarks form a social transaction perspective", *Housing Studies*, 13(6), pp. 781–792.

Blöss, T. (1989), "Jeunes maghrébins des quartiers nord de Marseille, une générations charnière", *Les Annales de la Recherche Urbaine (Familles, générations, patrimoine)*, 41, pp. 59–66.

Bonvalet, C., Brun, J. and Ségaud, M. (2000), *Logement et habitat, bibliographie commentée*, Paris: La Documentation Française.

Boumaza, N., Bekkar, R. and Pinson, D. (1999), *Familles maghrébines en France, l'épreuve de la ville*, Paris: Universitaries de France.

Brubaker, R. (1993), " De l'immigré au citoyen", Actes de la Recherche en Sciences Sociales, 99, pp. 3–25.

Brun, J. and Rhein, C. (1994), *La ségrégation dans la ville*, Paris: L'Harmattan.

Castells, M. (1972), *La question urbaine*, Paris: François Maspéro.

Chretien, J. P. and Prunier, G. (eds), *Les ethnies ont une histoire*, Paris: Karthala, ACCT.

Coing, H. (1966), *Rénovation urbaine et changement social*, Paris: Editions Ouvrières.

Delarue, J. M. (1991), *Banlieues en difficulté, la relégation*, Paris: Syros Alternatives.

De Rudder, V., Poiret, C. and Vorc'h, F. (1997), *La prévention de la discrimination raciale, de la xénophobie et la promotion de l'égalité de traitement dans l'entreprise. Une étude de cas en France – Rapport pour la Fondation Européenne pour l'Amélioration des Conditions de Vie et de Travail*, Paris: Université de Paris 7 et Paris 8, Unité de Recherches Migrations et Société.

Feldblum, M. (1993), "Paradoxes of ethnic policies: the case of Franco Maghrebins in France", *Ethnic and Racial Studies*, 16(1) pp. 52–74.

Hannerz, U. (1983), *Explorer la ville*, Paris: Editions de Minuit, collection Le Sens commun.

Haumont, N. (1996), *La ville: agrégations et ségrégation sociale*, Paris: L'Harmattan.

Horowitz, D. L. and Noigiel, G. (eds) (1992), *Immigrants in two democracies: French and American experience*, New York: University Press.

Kaufman, J. C. (1983), *La vie en HLM. Usages et conflits*, Paris: Les Editions Ouvrières.

Labat, C. (ed.) (1994), *Cultures croisées: du contact à l'interaction*, Paris: L'Harmattan.

Lefebvre, H. (1968), *Le droit à la ville*, Paris: Anthropos.

Martiniello, M. (1995), *L'ethnicité dans les sciences sociales contemporaines*, Paris: Presses Universitaries de France, collection Que sais-je, 2997.

Paugam, S. (1995), "L'habitat socialement disqualifié", in Ascher, F. (ed.) *Le logement en questions*, Paris: Editions de l'Aube, collection Essais.

Peraldi, M., Mozere, L. and Rey, H. (1999), *Intelligence des banlieues*, Paris: Editions de L'Aube, Territoires.

Petonnet, C. (1992), *Espaces habités, ethnologie des banlieues*, Paris: Galilée.

Pinson, D. (1992), *Des banlieues et des villes, dérive et eurocompétition*, Paris: Les Editions Ouvrières.

Poutignat, P. and Streiff-Fenart, J. (1999), *Théories de l'ethnicité*, Paris: Presses Universitaries de France, collection Le Sociologue.

Sarnad, D. J. (1978), "From immigrants to ethnics: towards a new theory of ethnicization", *Ethnicity*, (5).

Semprini, A. (1997), *Le multiculturalisme*, Paris: Presses Universitaries de France, collection Que sais-je, 3236.

Wirth, L. (1938), "Urbanism as a way of life", *American Journal of Sociology*, 44, pp. 1–24.

Chapter 3

The "Citadel of Exclusion": Regeneration Processes in the Area of Santa Maria della Pietà in Rome[1]

SILVIA MACCHI

The Regeneration of Derelict Areas

My work focuses on the regeneration of a particular type of urban area: specifically the areas, which housed the former psychiatric hospitals. These are part of the many urban structures inherited from our recent past, which have difficulty in rediscovering a functional role or an economically viable future in the context of the contemporary city. They can therefore be lumped together with a series of other structural components of the modern city (such as military barracks, factories, train stations and docks) which are now completely abandoned or under-used. Or they can be assigned to the wider category of derelict areas. Contemporary urban planning has devoted exhaustive attention to the problem of the re-conversion of derelict areas, in the attempt to reintegrate them in the processes of economic production and/or restore to them a socially useful role.[2] Such areas, however, have also attracted attention due to the urban movements they have generated and the "alternative" character of the projects for their regeneration that these movements have in turn formulated and supported.

The history of all derelict areas is, from this point of view, somewhat similar. In almost all cases it seems to me possible to identify in them a similar process of evolution, divided into three consecutive phases: abandonment, invasion renewal.

In the first stage we see the decline of the function for which the structure was originally conceived. The structure is then gradually abandoned by its owners (public or private). The duration of this phase depends on the degree of resistance mounted by those who worked in the structure, or who received from it a more or less vital social service. This resistance is almost always overcome in time. It's enough to sit back and wait for the former employees in the structure to reach retirement age and its social beneficiaries to grow old or die; the number and power of the people opposed to the abandonment of the structure is, in any case, drastically reduced by the mere passage of time. In some cases this process of wearing down the resistance of former workers or beneficiaries is accelerated by specific measures: early retirement, unemployment benefit, professional re-qualification, transfer to alternative structures etc.

As the "traditional" activities of the structure decrease, so the parts of it that have been abandoned are invaded by a new series of activities which, for various reasons, can be classified as "fringe" activities. They are, in fact, almost invariably located on the fringes of the economic and social process, and sometimes even excluded from it, because society does not recognize their economic value or social use. These activities find their ideal location in derelict areas, because it is just in such areas that they do not encounter the opposition of other interests stronger than they either in economic power or social legitimacy. In some cases it is the owners themselves who occupy, or who authorize the occupation of, these areas and buildings with facilities that take up a lot of space (archives, warehouses, repair workshops, etc.), or that need little or no new investment (such as parking lots or small-scale agricultural activities), or with temporary activities of any kind. In other cases the invasion stage is carried out by individuals or groups of citizens who find in derelict areas the spaces and buildings, which they have been denied elsewhere. This leads to the creation of playgrounds, football pitches, theatre workshops, market gardens, community centers, camp sites for travelling people, communal residences etc.

These various activities are usually tolerated even after abandonment has been completed. But they are often a source of conflict once the regeneration stage has begun. This stage, however, only starts once the surrounding conditions determine a new situation favorable to the economic exploitation of the property or its re-conversion to other uses. Usually a set of conditions has to be met before any regeneration is contemplated: the availability of public finance to convert the specific activity, the provision of new infrastructures for the area in question, the increased demand for building sites in that particular urban sector, etc. It is during this stage that the conflict between the owners of the site and the social groups that have invaded it becomes most pronounced. The outcome of this conflict – and particularly the probability of the social groups involved exerting any influence on plans drawn up for the re-conversion of the derelict area – greatly depends, however, on what happened during the preceding stages (abandonment and invasion). Have the invading groups developed any ties with the groups, which tried to resist the area's abandonment? Have they succeeded in co-operating together to produce a joint and (in cultural terms) widely shared project? Is this project able to mobilize the majority of the inhabitants of the area of the city in question? To what extent are the social groups involved in the project represented within the city's political institutions? What is the strength and extension of the social network that the groups involved have succeeded in developing around their project?

These are the questions that will guide my investigation of the particular case being studied here: the former psychiatric hospital of Rome, Santa Maria della Pietà (SMP). The decision to work on this particular derelict area was prompted by a hypothesis that I will try to develop in this chapter. In my view, the former psychiatric hospitals represent contexts particularly favorable to the establishment of urban communities (in the sense of aggregates of citizens united by the same ideal of urban life) endowed with their own autonomy of action and a certain political power.[3] This happens because the local process of abandonment, invasion and renewal derives its origin from a change in attitude, more particularly society's attitude, to madness, which contains the germs of a new "ideal of city life".[4] The outcome of the process depends, in this case, on the ability of the local population to embrace this new ideal and to put it into practice. Very different, it seems to me, is the case of industrial areas which have declined or been abandoned because they are unable to satisfy the new economic organization and post-industrial needs of our time. Here, too, a change in attitude is detectable, but this change has no intrinsic "positive" proposal for the local population. The new ideal of urban life needs, in this case, to be constructed *ex novo* and the transition from an attitude of resistance to one of insurgence (or even mere transgression) requires a great deal more energy.

The Decommissioning of the Provincial Psychiatric Hospital of Rome

The history of the decommissioning of the provincial psychiatric hospital of Rome began in 1978 with the promulgation of a new law concerning the reform of Italy's national psychiatric social services (Law no. 180/78: the so-called Basaglia Law). The law made provision for the dismantling of the big provincial psychiatric hospitals and their replacement by a network of territorial services for mental health-care. The Basaglia Law represents the institutional outcome of a process of psychiatric reform, which began in the sixties and whose main proponents were Ronald D. Laing in Great Britain and Franco Basaglia[5] in Italy. What these two authors challenged, above all, was the repressive and coercive nature of traditional psychiatry, as exemplified by the nineteenth century type of lunatic asylum.

The 1978 Basaglia Law was preceded in Italy:

- by the introduction of "voluntary admission" in 1968, in replacement of the "compulsory admission" prescribed by the previous legislation;
- by the experimental research conducted by Basaglia in Gorizia and Trieste since the early Seventies, with the aim of restoring civil rights to psychiatric patients and improving the quality of their lives (emblematic was the adoption of an "open door" policy for mental hospitals);
- by the setting up of *Psichiatria Democratica* in Bologna in 1973 (an association of psychiatric workers which would campaign for the closing down of mental hospitals and the reorganization of psychiatric services according to a logic of social inclusion instead of social exclusion).

The provincial psychiatric hospital of Santa Maria della Pietà (SMP) in Rome was totally unprepared for the 1978 reform. Little or no experimentation in that direction had been undertaken either inside or outside the hospital; no specific funds and no regional legislation existed for this specific purpose; and the available personnel were too few and too ill prepared for the reform. It was not until 1981 that admission to the hospital was finally discontinued. At much the same time Dr. Paolo Algranati, a young graduate doctor, began work in ward 22 (114 patients, one doctor, one ward sister, 30 nurses, locked doors and barred windows), but had to wait until 1983 before he was able to create a "free zone" for 13 patients and 6 nurses.[6] In 1978 about 1000 patients were being treated in the SMP; in 1995, almost 20 years after the Basaglia Law (1978), 380 patients still remained there. Given that the annual mortality rate among the patients of the former psychiatric hospitals was 5-6%, it seems clear that the other 620 patients were not "freed" but rather died from old age or disease.

The dismantling of the psychiatric hospitals underwent some acceleration in the mid-nineties. In 1994 the government issued a decree (DPR 07/04/94) specifying a series of specific measures aimed at the rapid phasing out of the system of psychiatric hospitals. A few months later Law 724/94 set 31 December 1996 as the deadline for the closure of these hospitals. Two years later, just before this date came into effect, another law (662/96) confirmed their compulsory closure and defined the economic sanctions to be paid by those Regions that failed to comply with the so-called "plans to replace the psychiatric hospitals" by 31 January 1997. Finally, in 1998, further legislation was issued, prohibiting the reutilization of the areas and buildings of the former psychiatric hospitals for new mental health services. Once their existing functions had ceased, these areas and buildings, if they could not be used for non-psychiatric health services, were to be used to generate income, through sale, rent or mortgage, with the aim of raising the necessary funds to finance the new mental health services as defined by the Basaglia Law of 1978. The combination of these provisions forms part of a broader program aimed at increasing the managerial efficiency and financial accountability of the Italian public services; various governments have promoted this change in direction since the early nineties to help balance the budget for public finance.[7] Thanks to these provisions and the economic sanctions to be paid by the institutions not in compliance, the number of patients being treated in the SMP had decreased to less than 300 in 1996. While awaiting the new regional structures, these patients have been organized in autonomous psychiatric units, which, apart from their location, have nothing in common with the old model of psychiatric hospitals.

In the meantime, what happened to the buildings and open areas thus gradually freed from their original function? In 1978 the SMP, in structure and appearance, remained much as it had been conceived and realized in the early years of this century: a small garden city comprising about 30 buildings (*c.*200,000 cu m) distributed throughout an extensive area of parkland densely planted with trees (26 ha). To this we should add the adjacent farm which allowed the SMP to be completely self-reliant in food supply and which was also used for work therapy for the patients. On the other hand, the urban context in which the SMP is situated has undergone a great deal of change. At the time of its foundation, the SMP was located in the open countryside, even though only four kilometers as the crow flies from the Vatican. In 1978, by contrast, the SMP was situated within the built-up area of the city of Rome. It was surrounded on all sides by dormitory quarters of low-income settlements and illegal private housing which had sprung up in an unplanned fashion in the post-war period, without adequate social services and transport infrastructures.

In 1978 the history of the hospital and that of its agricultural estate began to diverge, due to a series of factors. First, their respective ownership changed: the hospital was assigned to the public-sector company (ASL Rome E) responsible for managing the health services in the northwestern sector of Rome on behalf of the Region; the farm remained the property of the Province. Second, the necessary time-scales for the interruption of psychiatric activity were different: whereas the farm could be abandoned rapidly, the abandonment of the hospital required that new structures first be created for the patients and personnel. And third, different economic and symbolic value attached to the two structures: whereas the hospital comprises an important complex of buildings and parkland, the agricultural estate is an open area of little aesthetic and natural beauty, whose value is only that of its potential use as a building lot.

Nonetheless, still today it is impossible to predict whether the two structures will have a different future or whether they are destined to be reunified in a single plan of urban regeneration. In fact, their destiny is inseparably linked with that of the urban sector to which they belong. Hitherto any plan put forward by the owners for their economic exploitation has been thwarted by the lack of good road and rail links with the rest of the city and by the general conditions of degradation of the whole urban sector. Moreover, it should not be forgotten that the hospital and its agricultural estate have always been destined by the town plan of Rome to the realization of public services of general interest (hospitals, universities, prisons, etc.). Any project at variance with these aims would thus require a change in the town plan, which can only take place with broad public support. Lastly, a part of the agricultural estate was declared a protected green belt by the last municipal administration headed by a "green" mayor. Consequently, the possibilities of economic exploitation of the estate have been further reduced over the last twenty years, without any substantial improvement being achieved in access to the area.

The history of the hospital and that of the agricultural estate since 1978 can be briefly summed up in terms of the tripartite scheme we proposed above: abandonment, invasion, renewal.

a) The Hospital

- the abandonment stage began in 1981 when new admissions were discontinued and ended in 1996 with the termination of the "regime" of psychiatric hospitals. The main protagonists of this phase were the former psychiatric patients and staff;
- the invasion stage began in 1981 with the opening of the area to the public as a park and is still continuing in the early months of 1999. One of the pavilions is at present occupied by squatters. The protagonists of this phase are the inhabitants of the neighboring quarters, some co-operatives of socially useful work set up by former psychiatric patients and personnel, and numerous associations – local, metropolitan or national – interested in producing "alternative" culture (theatre, painting, etc.) or defending the natural environment;
- the regeneration stage began in 1996 when the first proposals for alternative use were made. Most of the associations present within the former hospital and in its neighboring quarters have set up a joint committee (Coordinamento "Città Ideale").[8] Its objective is to ensure that the associations of citizens participate in the decision-making process and have an influence on it equivalent to that of the owners.

As things stand today, no final project as yet exists, and formal negotiations between the various interested groups have yet to be initiated. The first structural modifications have, however, been undertaken to transform some of the former hospital pavilions (9 out of 33) into temporary accommodation for the pilgrims expected for the Jubilee Year.[9] Out of the remaining 24 pavilions, 11 are abandoned and 13 are utilized as follows: eight for specialized or community non-psychiatric health services, two for non-profit activities, two for temporary accommodation for ex-psychiatric patients, one for lodgings for squatters.

b) The Agricultural Estate

- the abandonment stage was apparently very rapid, since every activity connected with the psychiatric hospital ceased in 1980 when the properties were split up. But the former farm workers on the estate still live in cottages on the estate, even though they no longer work on the farm or even for the Province;
- the invasion stage began in 1981 and is still continuing today for more than three-quarters of the area. The protagonists of this phase are: a farm co-operative set up to provide a source of employment for former psychiatric patients and youths detained in the nearby juvenile remand center; the inhabitants of the surrounding area interested in the cultivation of small market-gardens; some sports associations; a community of travelling people; and a few small private enterprises (nurseries, sheep-farming, horse-breeding, etc.);
- the regeneration stage began in the early eighties, with the opening of three schools and some sports facilities, but is proceeding slowly and in an unplanned fashion. The protagonists of this phase are still confined to the public administration alone (that of the city, region and government ministries, etc.), which in general limits itself to requesting the owners for the concession of small areas for the realization of public services.

Hitherto, neither the owners nor the citizens have committed themselves to the formulation of a comprehensive plan for the whole estate. Only in the last three years has there emerged a ground swell of public support for a large part of the estate being maintained as a green belt. The present owners, by contrast, seem only to be waiting for a suitable occasion to sell the estate as a building site. The result is a situation of general degradation, with a myriad different activities (legal and illegal) cohabiting together in the midst of mounds of refuse and drainage ditches transformed into open sewers.

A New Attitude to Madness: from Internment to Contamination

As I have already said, the interest of this case study consists in the particular change in attitude to mental illness, which gave rise to the process of abandonment, invasion and renewal we have described. This change in attitude regards in the first place the SMP psychiatric community, i.e. the sum of the patients and psychiatric workers (doctors, nurses and other staff) who were living and working in the SMP in 1980. In my view, this community had the merit of responding positively to the stimulus of psychiatric reform and transmitting it to others in the form of a new approach to community life. A fundamental stage in this process was the transformation of the psychiatric community from a numerical into a political minority. At the time of the Basaglia law the psychiatric community was a minority only in the strictly numerical sense: because of its social isolation, it had no identity at all at the political level. Organizing itself as a minority group (a social group without direct decision-making power but with the ability to voice its opinions to the relevant city authorities) took a great deal of work: psychiatric workers and patients had not only to change the relationships existing inside their community, but to build up a social relationship with the outside world in order to break out of a chronic situation of social exclusion. In my opinion, the basic precondition for the success of this task may be identified in the dissemination of a new attitude to mental illness.

This radical change in attitude permitted this community not only to overcome its own social exclusion, but to go well beyond it. The patients and staff of the SMP were thus able, by their work, to give rise to a wider movement against social exclusion which has involved, in different ways, the inhabitants of the surrounding area, the city of Rome and a series of territorial contexts, some even far away. Not only has the new attitude to mental illness now been accepted by many inhabitants in the area and by some civic political groups, but the SMP has become a symbol of the struggle against social exclusion. Thanks to its new identity as "citadel of exclusion", a place where social outcasts can rediscover their lost voice, the SMP is now able to act as a magnet for movements active even in fields far removed from psychiatry but similarly involved in the campaign for civil rights. The loose alliance of pressure groups thus formed (spearhead of an ever wider network) has demonstrated its ability to oppose in an effective manner the various plans for the "pure" economic exploitation of the former psychiatric hospital formulated by its owners and, even if it has not succeeded in getting its own way, has at least managed to steer the decision-making process towards objectives more in line with its own social agenda. In fact, as yet no final plan has been adopted for the SMP, but undoubtedly, thanks to the pressure exerted, some solutions that would have wiped out the last twenty years of history have been excluded. In particular three different plans put forward by the owners have been jettisoned: the establishment of a university which would have occupied the whole hospital and led to a large part of the agricultural estate being built over; the transformation of the former hospital into an administrative center, with private offices and related commercial services; and the re-conversion of the hospital into film studios, with its own multi-cinema complex.

But what are the constituent elements of this new attitude to madness? The history of "madness", and the derivation of the modern concept of the "mentally ill", have been extensively analyzed by Michel Foucault in *Histoire de la folie à l'âge classique* (1961). In this, as in other of works of the French philosopher, we may find many arguments and ideas in support of our reflections on the new attitude to madness.

Foucault distinguishes between two periods in the history of madness: a classical period, which has its roots in the Renaissance and was fully manifested from the seventeenth century by the establishment of "madhouses" for the isolation of the mentally ill throughout Europe; and a second period, which we could define as that of the Enlightenment, which began towards the close of the eighteenth century and would lead in the following two centuries to the creation of the "modern" insane asylum. To these two phases we can add a third; a phase in which more recent advances in neurosurgery and psychoanalysis variously contributed to consolidating the barrier between society and the insane, the foundations of which had been laid in the Renaissance.

According to Foucault, the Renaissance marked the end of the mystic, or religious, view of insanity, as also of poverty in general. After the Reformation, the poor man was no longer considered the representative of God on earth, one of "the last who shall be first" of the Gospels, but an individual predestined to carry the burden of divine punishment. The poor man who rebelled against his condition sowed the seeds of social disorder; he became a beggar and a vagabond, wandering through the city in search of alms. So places of internment were created to defend the social order (outside) and to correct the anti-social behavior of the poor (inside). Citizens were urged to re-direct their charity towards these institutions, since the giving of alms directly to the poor was condemned as a way of reinforcing or perpetuating their anti-social behavior. So began the age of "mass internment", which would lead in the seventeenth century to one Parisian out of every hundred being locked up for various reasons.

Not until the Enlightenment, however, did madness come to be distinguished from poverty, lose its connotation of "divine punishment", and be labeled instead as "mental illness". But that does not mean that the internment of the mentally ill had ceased. At the end of the eighteenth century, Philippe Pinel (1809) in France conceived the first modern "insane asylum"; he transferred to it the mentally ill, who had hitherto been indiscriminately kept behind bars together with other socially dangerous categories. By this gesture modern psychiatry defined its object of study. During the Enlightenment, the mentally ill person was seen not as someone who had lost his reason, but as someone who was suffering from a disorder of the spirit which originated from contact with unhealthy social contexts in both the moral and the physical sense. He was considered curable. His treatment was based essentially on two therapeutic principles: first, isolating him from the origin of his malady by placing him in a morally and physically wholesome environment (the modern insane asylum); second, inculcating him with a consciousness that he himself was responsible for his own insanity and punishing him whenever he lost control of himself. While madness itself was no longer considered a crime, the madman was considered guilty, and hence punishable, if he did not make an effort to get well. The chains and bars had been eliminated, but what remained was the need for society to be protected from the excesses of madness, and the insane to be protected from the depravations of society. The insane asylum was no longer a place of terror: on the contrary, it was conceived as a comfortable and "humane" environment where the insane could move about freely (unless it was necessary to punish them), though it remained surrounded by a wall to prevent any risk of contamination.

With the rise of positivism and advances in medical knowledge, the relation between mental derangement and an "insane" (= unhealthy) context would increasingly lose its social connotation and come to be seen in purely biological terms. By the end of the nineteenth century madness was no longer regarded as a temporary disturbance of the spirit but rather as the symptom of some physiological dysfunction in the brain, caused by depraved behavior either by the individual himself or by his parents (alcoholism, sexual promiscuity etc.). This concept, which would provide theoretical justification for terrible eugenic experiments, would also give rise to new therapeutic techniques (neuro surgery, electric shock treatment, pyretotherapy etc.) that concentrated wholly on the body and reduced the "moral" (or reformatory) treatment practiced at the beginning of the century to at best a marginal role. At the same time, patient internment would lose its function of protecting the patient and be reduced to a mere instrument of social defense. Not even psychoanalysis, which would contribute decisively to discrediting the dangerous determinism of early neuropsychiatry, would be able to create a breach in the high walls that surrounded the mental hospital: indeed, it would only reinforce further the social isolation of the mental patient and his doctor.

The new attitude to mental illness that the SMP psychiatric community succeeded in introducing is aimed, in the first place, at breaking down the barriers – both concrete and symbolic – that surround the mentally ill and that isolate them from society. The overriding objective is to restore civil rights to psychiatric patients. The main means of achieving this end is that of mutual contamination: the contamination of society by mental illness and of mental illness by society. This strategy is summed up by the SMP slogan *"entrare fuori per uscire dentro"* (to come in from outside in order to go out from inside). Ever since the early eighties, in fact, the doors of the SMP have been open both for the public to enter and for the patients to leave. The patients thus began to refamiliarize themselves with the outside world: they began to go into the city, first with brief outings (excursions, visits to the cinema, to cafés, etc.) and then with ever greater regularity (often to go to work), until many of them moved out of the SMP altogether (with their health workers when necessary), either to take up residence in apartments or small houses in the suburbs or even to return to their hometowns. At the same time, the populace at large was encouraged to enter the SMP park and to regard it as they would any other civic park. In this way they began to get interested and involved in the patients' initiatives (exhibitions of paintings and ceramics, for example) and eventually to co-operate with them in organizing other ventures (such as theatre workshops, film festivals, concerts and conferences) until the terms "inside" and "outside" had lost much of their meaning.

In 20 years of hard work, the SMP psychiatric community has succeeded in establishing a rapport between its own mental illness and the citizens of Rome. As in all relationships, there have been moments of crisis, break-ups and steps backwards. There is no lack of protests against the "crazies" who wander freely about the city disturbing the peace. But what no longer exists is the insane asylum as a solution: the physical segregation of the insane by immuring them in a psychiatric hospital. The citizens of Rome have now learned that they cannot avoid cohabitation with mental illness; that it can no longer be hidden away behind a wall; that day after day they must live with the mental derangement in their midst, wherever it manifests itself. If "secure" means "without care, without concern",[10] the Romans today realize that there is no security against madness in the absence of care, that they cannot liberate themselves from it, and that it can only be coped with on a daily basis by constant care and concern.

The SMP psychiatric community has succeeded in making others "appreciate" madness, whether they are those closest to it in geographical situation or in political sympathy. Those who have had daily contact with the former patients of the psychiatric hospital, whether out of choice or duty, have learned to look beyond the illness and recognize the common humanity hidden by the mask of madness. From this point of view, the creation of art studios and the exhibition to the public of the patients' works have perhaps been the most effective means of bringing about inter-communication.[11] If, as Foucault writes (1961), madness is the *absence of work* and, conversely, *"là où il y a oeuvre, il n'y a pas folie"*, then the creative works of the psychiatric patients are a tangible demonstration that they are not only "madmen"; that they are not something different from any other human being; and, above all, that they lack nothing to be able to be human beings in the full sense. Foucault, however, enables us to go beyond the simple antithesis between "madness" and *"oeuvre"*, and provides us with conceptual tools to understand the complex relation between them. Madness interrupts, but does not destroy, the *oeuvre*: through madness, the *oeuvre* generates a void, a moment of silence, an unanswered question which is imposed on the rest of the world and causes it embarrassment; it unsettles it, it makes it feel guilty. The silence of madness is the cry of a man who has found no one to listen to him, it is the supreme expression of the revolt expressed by the *oeuvre* itself:[12] the extreme denunciation of the responsibilities of a world which the *oeuvre* calls into question.

> Cunning and new triumph of madness: this world that thinks it can measure madness, and justify it with psychology, must on the contrary justify itself before it, because, in its effort and in its disputes, it has to measure itself against the immeasurable dimension of works such as those of Nietzsche, Van Gogh and Artaud. And nothing in that world, least of all what it can know of madness, can make it certain that these works of madness justify it.

With these words, Michel Foucault concludes his work on madness.

The SMP as a Focus for the Prevention of Exclusion, Psychiatric or Not

The SMP today is an integral part of the city, a place where the inhabitants of the neighborhood bring their children to play and walk their dogs, a place where the elderly meet for a card game and the young come to jog. Its normality, however, is also pervaded by a quality of "specialness": it preserves the memory of exclusion and mental suffering and the struggle against both.

The graffiti of former patients are preserved on the walls of the "abandoned" pavilions, while the "murales" of the new citizens adorn the walls of the "invaded" pavilions. Next to the remains of the old fenced-in enclosures, which surrounded the inmates when they came out for a breath of fresh air, are the little orchards and greenhouses which later invaded the area. The SMP's official archive, comprising the documentation of 450 years of psychiatric treatment in Rome (not only patients' files, but also the beds in which patients were forcibly strapped down and other "therapeutic instruments"), is still incomplete: it still remains to be combined with the documents produced by the psychiatric community of the SMP in the last twenty years (again patients' files, but especially short stories, poetry, paintings and drawings, which express both the voice of the psychiatric staff and of the patients).

But the historical legacy preserved in the SMP today does not consist merely of documents or objects. For many of the protagonists of that history still live and work in the park, in the former hospital pavilions and in the various areas of the agricultural estate. A co-operative of former patients is responsible for cleaning the premises used by the health services. Other co-operatives, often with the participation of former health workers, run small handicraft workshops. Some of the former patients who, because of advanced age or physical handicap (brain damage), cannot be incorporated in the new psychiatric services, reside permanently in the SMP. Apart from the psychiatric community itself, a series of other activities has occupied the SMP. They are activities that reflect the new attitude to mental illness or, more generally, the theme of the quality of life: therapy groups for the prevention of learning and language disabilities in children, programs to develop multi-media technologies for social objectives, environmental projects, courses on solar technologies, etc. Even the camp for travelling people, which occupies a small part of the former agricultural estate, has felt the benefits of the presence of the SMP; thanks to the political support of those social groups that interpret the former insane asylum as a focus for the struggle against exclusion, the travelling people of the SMP have succeeded – a case unprecedented in Rome – in obtaining a regular contract for the supply of electricity without having to succumb to the standard blackmail of "normalization" (civil registration, expulsion for members of the group without residence permits, etc.).

This legacy of objects, people and activities is now at the center of the debate concerning the future of the SMP.

As already mentioned, the law requires the public-sector health service company that now owns the SMP (ASL Rome E) to proceed to the economic exploitation, i.e. profitable sale, of the real estate comprised by the former psychiatric hospitals and allocate the funds thus raised to the provision of new mental-health services. It is obvious, however, that any measure taken with profit as its sole objective would be incompatible with the conservation of the more intangible legacy described above. In strictly economic terms, indeed, the real estate is devalued by the presence of people, things and ideas connected to the remote and more recent history of the SMP.

On the other hand, the various social groups, local or not, psychiatric or not, that recognize this legacy as of irreplaceable human, political and cultural value have created the conditions in the course of time to oppose the owners' plans. They have succeeded in creating a network of individuals and groups that operate on various levels: in the offices of the public administration and in the seats of political debate, in the social venues of the local community and in the cultural centers of the city. Their message is simple: if the only reason to proceed to the economic exploitation of the SMP is to finance the new services for the safeguard of mental health, why not recognize that the SMP as it exists today is already a concrete realization of this objective? Is not the struggle against exclusion the fundamental prerequisite for the protection of mental health? And what is the SMP today if not the symbolic place of this struggle, the place from which social exclusion seems to have been effectively banished.

In effect, the existing legislation requires that the properties of the former psychiatric hospitals be used to generate income only if they cannot be directly used for the purposes of health care. At this point the question is posed, to what extent do projects aimed at improving the quality of life in a broad sense fall within the scope of the health service management. In other words, the SMP's social groups are making the same challenge to the health service management that was made 20 years ago by the psychiatric community: i.e. basing plans for the future of the SMP on a fundamental change in attitude.

To complete the picture of the situation, it should be recalled that the former psychiatric hospital is subject to a whole series of town planning restrictions which need to be observed and would require the full political consent of the organs of municipal government if they were ever to be overridden. The restrictions in questions prescribe, first, that the entire area shall be destined for "general public services"; second, that any plans for the transformation of the site shall not alter the volume and shape of the existing buildings, nor damage the trees in the park; and, third, that the area of the former agricultural estate that has not yet been redeveloped shall remain a green belt. The erection of new buildings both within the former psychiatric hospital and in its adjacent areas is thus precluded, and whoever acquires the existing buildings shall be under an obligation to assume responsibility for the maintenance of the park.

At this point, the owners of the SMP find themselves squeezed between two opposing forces: on the one hand, the pressure of the social groups that are able to enlist the support of private citizens, interest groups and political parties; and on the other, the power of the municipal administration, which is little inclined to relinquish its prerogatives unless it be for a project that doesn't create too many complications. This conflict is certainly not favorable to a process of rapid and economically efficient urban renewal. After four years of negotiation, restructuring has begun on some of the former hospital pavilions and on part of the park with funds destined for the Jubilee. But the conditions imposed on the properties have postponed any possibility of treating the SMP as a purely economic concern for at least ten years:

- the renovated pavilions shall be used for the accommodation of pilgrims with low income and/or health problems;
- these same pavilions shall be managed for 10 years by social co-operatives that guarantee employment to disables people, ex-prisoners etc.;
- the park shall be equipped with the necessary facilities to permit its use as a public park, and the more urgent measures be taken for the conservation of its trees;
- the management and upkeep of the park shall be assigned to the city park services (also for 10 years).

In the meantime, the city authorities have taken steps to draw up a plan for the regeneration of the whole district and to carry out the most urgently needed infrastructural works. In this regard, a series of measures aimed at improving road and rail links are in the process of being completed; they should considerably improve access to the area and to the SMP in particular.

Measuring the "Success" of the SMP

The history of the SMP over the last 20 years has its origins elsewhere: i.e. in the wider movement which, thanks to Basaglia and Laing, campaigned in the sixties and seventies to change the approach to psychiatry. This, to a certain extent, holds true for all derelict areas, in the sense that the cause of their rundown is very rarely endogenous. What, however, distinguishes the areas of the former mental hospitals from other derelict areas is the objective of the movement which brought about their demise. The whole psychiatric reform movement was aimed at improving the living conditions of the mentally ill by providing for their basic human needs. The industrial restructuring, which formed the root cause of other closures, was aimed, by contrast, at increasing profits. In the first case, the former patients and staff are the protagonists of change; in the second case, the former workers or beneficiaries are no more than a force of inertia, resisting change.

The merit of the psychiatric community of the SMP, however, was to have appropriated the dynamics of change in a constructive way and thereby transformed the process of psychiatric reform into a small urban revolution. The decision to "come in from outside in order to go out from inside" implied that the contribution of "outsiders" (inhabitants of the neighborhood etc.) be placed at the center of the SMP's reorganization process. Even though the relationship between local government and SMP community has not always been ideal, the local authorities have never been able to ignore or openly oppose a movement that included such a wide and varied range of social groups.

Today, those who suffer from mental disabilities know that there are those who will take up the battle on their behalf: those who will give their silence a voice and bring their suffering protests to the forums of political debate. These spokesmen of mental illness, like all other envoys, will always incur the suspicion of having betrayed their mandate. Who, in fact, exercises control over them? Who checks to see whether their translation of the silence of the mentally ill into political representation is faithful or not? The success of the SMP project (and of the network of movements it includes) must, in my opinion, be measured by the extent to which an answer to these questions is found. If the SMP does not assume responsibility itself for protecting the anti-exclusion movement from possible betrayal, the risk is that the power of its "spokesmen" will increase without any benefits for those they represent.

The SMP, however, is very conscious of this risk. Right from the start, there have been those who have exploited former patients to obtain economic benefits of various types (low-rent or even rent-free property, tax reductions, cut-rate credits, etc.), and then sidelined the former patients or even ousted them altogether from their activities. The law that made provision for the expropriation of the former psychiatric hospitals itself laid itself open to various abuses of this kind. Its primary objective was in fact the improvement of public finance; it therefore defined very clearly the terms under which the psychiatric hospitals should be closed and the procedures for the economic exploitation of their accompanying real estate. It said a good deal less about what was to become of the former patients and how they were to be assured a really alternative future in terms of quality of life. Many former patients have been transferred from the state psychiatric hospital to private nursing homes, and it is to these latter (a sort of "private mental hospital") that the families of new psychiatric patients are obliged to turn today. In fact, many people today accuse the State of having concealed the privatization of the psychiatric services behind the specious claim of a sacrosanct mission: the accelerated implementation of the reform inaugurated in 1978.

The monitoring of the work of the spokesmen, however, still hasn't gone far enough. If we are really willing to recognize in the elderly person who is mentally ill, someone who has opted for silence as an extreme form of protest, what right do we have to force him/her to speak and, what is more, to speak in political terms? As Isabelle Stengers points out in a recent essay (1997), the problem of translating any form of social disability into political terms is a problem of Western democracy. Western democracy functions like this; the achievement of social peace is based on the assumption that each citizens wishes or is able to identify himself with some collective political platform. The insane, like all those who belong to "different" cultures or who refuse representation, alert us to the need to find alternative ways of achieving social peace. We need to go beyond politics and enter the uncharted terrain of "cosmopolitics"[13] or the "politics of difference".[14] But, in our case, a precondition for this to take place is that the SMP be a citadel not only for those who campaign against exclusion (the "spokesmen"), but also for those who accept the exclusion of their condition and silently experience it in their daily lives (including the insane, in spite of the prescriptions of the law).

Notes

1   This chapter was written in 1999. The editorial decision has been made to retain the temporal references it contains as these do not detract from its analysis.
2   One of the few texts that throws doubt on this type of logic with specific application to town planning is Lynch (1990).
3   A complete documentation on the present state of these areas in Italy has been assembled as part of a specific research project conducted by the Fondazione Benetton Studi Ricerche (Luciani et al., 1999).
4   According to Iris Marion Young (1990), an "ideal of city life" is a vision of social relations as an affirmation of group difference.
5   The writings of Franco Basaglia are collected in two posthumously published volumes (Basaglia, 1981 and 1982).
6   In a recent book, Paolo Algranati (1999) tells about his 20 years of work at SMP. Today his patients live in the Peter Pan community (http://www.geocities.com/Paris/4179/index.htm).
7   This seems to be the real reason why various governments pressed for the speedy closure of the psychiatric hospitals, rather than any concern for a rapid social reintegration of psychiatric patients. If we compare the dates of the above-cited legislative provisions with those of the periods in office of successive Italian governments, we may note a substantial continuity in psychiatric policy, in spite of the fact that there was an alternation between center-left and center-right governments. In particular, Law 724/94 was approved during the Berlusconi government, elected with the support of parties that have always campaigned against the Basaglia reform.
8   httm://www.geocities.com/CapitolHill/7170/cittaid.htm.

9   The year 2000 is being celebrated as a Jubilee year by the Catholic Church. It is predicted that over 30 million pilgrims will visit the city of Rome in the course of the year to participate in the religious ceremonies organized by the Holy See. The present mayor of Rome, in office since 1994, has placed this event at the center of his political agenda and has received from the national government the job of coordinating all the necessary building and infrastructural works to prepare the city for the holding of the Jubilee.

10 The term "secure" has a Latin etymology and derives from "se" (separate) + "cura" (care, concern).

11 The artistic talents of former SMP patients have achieved various recognition. The schools of the district regularly invite them to work alongside students in school workshops; some of their works have been put up for sale on the art market; and the media have on several occasions shown an interest in their work; for example, the painter Baieri, former SMP patient, became famous after participating in a popular TV talk-show.

12 The relationship between art and revolt is effectively expressed by Albert Camus in L'Homme Révolté (1951). This theme has been more recently taken up by another French philosopher, Michel Onfray (1997).

13 "Cosmopolique" is a term coined by Isabelle Stengers (1997) to define an area for thought that examines the relationships between science, technology and culture in an attempt to define an alternative for "political" peace.

14 In referring to the "politics of difference" we allude in particular to the theoretical work of Iris Marion Young (1990). Young yearns for a different kind of city that is organized not to suppress difference and oppress "others" (i.e. whoever differs from the predominant norm), but to assimilate them: a city, in other words, that is open to "unassimilated otherness" (cf. also McDowell, 1999).

References

Algranati, P. (1999), *Voci dal Silenzio*, Milano: Elèuthera.

Basaglia, F. (1981), *Scritti 1953–1968. Dalla psichiatria fenomenologica all'esperienza di Gorizia*, Torina: Einaudi.

Basaglia, F. (1982), *Scritti 1968-1980. Dall'apertura del manicomio alla nuvovo legge sull'assistenza psichiatrica*, Torino: Einaudi.

Camus, A. (1951), *L'homme révolté*, Paris: Gallimard.

Foucault, M. (1963), *Histoire de la folie à l'âge classique*, Paris: Plon.

Luciani, D. et al (1999), *Per un atlante degli ospedali psichiatrici in Italia*, Treviso: Fondazione Benetton Studi e Ricerche.

Lynch, K. (1990), *Wasting away*, San Francisco: Sierra Club Books.

McDowell, L. (1999), "City life and difference: negotiating diversity", in Allen, J., Massey, D. and Pryke, M. (eds), *Unsettling Cities*, London: Routledge, pp. 95-136.

Onfray, M. (1997), *Politique du rebelle. Traité de résistance et d'insoumission*, Paris: Grasset.

Pinel, P. (1809), *Traité médico-philosophique sur l'aliénation mentale*, Paris: J.A. Brosson.

Stengers, I. (1997), *Cosmopolitiques. Tome 7. Pour en finir avec la tolérance*, Paris: La Découverte/Les Empêcheurs de penser en rond.

Young, I. M. (1990), *Justice and politics of difference*, Princeton, NJ: Princeton University Press.

Chapter 4

Urban Transition and Immigrant Entrepreneurship: Processes of Creation of Openings for Immigrant Businesses in Amsterdam and Rotterdam

ROBERT KLOOSTERMAN AND JOANNE VAN DER LEUN

Immigrant Entrepreneurship and Economic Change[1]

At first, the appearances of the people in the streets of Dutch cities changed. Different faces, different fashions, and different languages. More recently, streets in many of these cities have undergone another remarkable outward transition. The products displayed in the shop-windows and, moreover, the entrepreneurs themselves are now reflecting the changes in the urban population due to large-scale immigration. Because of this, these cities have become increasingly cosmopolitan, both in the sense of having a more diverse population, but also in resembling more other advanced urban economies, such as New York, Los Angeles, London and Paris, where immigrants and immigrant entrepreneurs are prominently present.[2] As in these cities, a growing part of the population of Dutch cities originates directly from Third-World countries or has parents who were born abroad (SCP, 1998).

In terms of upward social mobility, setting up shop by immigrants could be interpreted as an attempt to pull themselves up by their own bootstraps. From a somewhat different perspective, they could be seen as local stakeholders who could help to initiate or strengthen regeneration processes in deprived city neighborhoods. By bringing back businesses in derelict shopping streets, by offering job opportunities, and by furnishing role models for their compatriots, vicious circles in neighborhoods could be stopped or even turned around (Kloosterman and van der Leun, 1999). Immigrant entrepreneurship, consequently, should be seen as a very important category of rather autonomous economic agents in urban economies.

The way they affect urban economies in general and neighborhoods in particular, however, is contingent on their opportunities to carve out a decent living as an entrepreneur. If immigrant businesses are destined to stay marginal affairs, their impact will be rather limited. Below, we explore the opportunities for immigrant businesses from a specific perspective that is informed by the more structural changes in advanced urban economies. We will, more particularly, examine how openings for immigrant businesses emerge and, furthermore, how this affects chances of immigrant entrepreneurs to expand their economic activities.

The rise of immigrant entrepreneurship, at first glance, seems the rather obvious outcome of, on the one hand, the presence of a large number of (aspiring) immigrant entrepreneurs, and, on the other, the rising number of opportunities for small businesses. The matching of supply and demand on the entrepreneurial market is however, anything but self-evident. The share of a particular group or category among the ranks of the self-employed is never a simple reflection of their share of the population as a whole. The rate of participation in entrepreneurship of a particular group of immigrants depends both on the socio-economic and cultural characteristics of the group in question (the supply of potential entrepreneurs) and on the particular shape of the *opportunity structure* for starting businesses: the "demand side" for entrepreneurs (cf. Engbersen and van der Leun, 1998).

Below, we distinguish two theoretical points of view that are derived from more general interpretations of the dynamics of advanced urban economies. The first point of view assumes that the opportunity structure changes in a polarized way creating openings both at the high and the low end of the opportunity structure. This latter expansion may benefit immigrant entrepreneurs. The second point of view does not take changes in the opportunity structure as its starting point. Instead, the emphasis falls on processes that resemble a game of musical chairs: openings for (recent) immigrant entrepreneurs are created when already longer-established entrepreneurs are able to secure a better living (either as a worker or still as self-employed) in a different setting. Thus, openings are provided at the lower end of the opportunity structure for newcomers who do not have much choice and still derive their preferences mainly from the sending country.

On the basis of data on starting entrepreneurs in Amsterdam and Rotterdam provided by the Chambers of Commerce, we will assess which of these two points of view offers the best fit with the recent rise of immigrant entrepreneurship as observed in the two most important Dutch cities. We start with a more general analysis of the matching process of supply and demand in the entrepreneurial market.

The Matching of (Immigrant) Entrepreneurs and the Opportunity Structure

The scope for starting a business by fledgling entrepreneurs in general is contingent on the so-called opportunity structure: the number and the nature of openings for small businesses. This opportunity structure – which in itself is primarily a function of the state of technology, the costs of production factors, the nature of product demand, the institutional make-up – determines where and to what extent the opportunities for such businesses will occur. In particular in the last quarter of this century, it seems that the viability of small-scale enterprise has strongly increased.

This increase of opportunities for small businesses in general is connected with several factors. Firstly, the development of small-scale information technological applications has enlarged the scope for small firms. Secondly, the increased tendency among larger businesses to contract various activities out, such as catering, security, cleaning etc., has meant more opportunities for small businesses. Thirdly, the fragmentation of particular product markets into all kinds of sub-markets with the result that larger series are not all that profitable anymore, thus eroding economies of scale, has changed the seemingly inevitable dominance of large firms

(OECD, 1993; Van Zanden, 1997)[3] Fourthly, changes in regulatory frameworks in advanced economies have opened up possibilities for small firms.[4] This is partly the result of a more general political program of liberalization, partly because of the widely held expectations that new firms would be able to alleviate the unemployment problem that plagued many advanced economies after 1980 and to strengthen the long-term potential of these economies.

The international renaissance of small-scale enterprise took place more or less synchronously with a new phase in immigration to advanced economies, namely in the second half of the 1970s. This new phase was characterized by shifts in existing migratory patterns and new forms of migration. These changes included a decline of labor migration to Western Europe, the development of mass movement of refugees and asylum seekers and a rise of immigrants due to family reunion of former foreign workers and colonial workers (Hollifield, 1992; Castles and Miller, 1993). These two latter migratory movements indicated – in contrast to the former labor migrants – a conscious choice by migrants for more permanent settlement in the host country. This change in the mental make up of immigrants from the 1970s onwards has obvious consequences for the amount of investment they are willing to make in their new country and, thus, also for their proclivity towards setting up their own shop.

Although the demographic side of the equation is met and although the opportunity structure has evolved in a way that is more favorable for small businesses, the rise of immigrant entrepreneurs from Third-World countries in advanced economies is, however, still anything but an obvious or necessary outcome. Complex and often selective processes of interaction take place between demographic developments and changes in the opportunity structure.

Firstly, shifts in the opportunity structure in advanced urban economies that increase the number of openings for small firms do not always have to concern activities that require only small outlays of capital, and involve only relatively low skills. Most immigrant entrepreneurs from less developed countries are dependent on just these kind of openings (Kloosterman, 1996a).

Secondly (potential) immigrant entrepreneurs have to perceive and subsequently seize opportunities that occur. If they do not show any significant proclivity towards entrepreneurial culture or if they lack the required skills and qualities for setting up businesses, opportunities might remain unused. In contrast to what neo-classic economics tends to assume, the supply of entrepreneurs is not a highly elastic, endogenous outcome of transparent economic processes (cf. Light and Rosenstein, 1995).

Thirdly, even if opportunities at the lower end occur and are, moreover, also perceived by those who want to seize them, immigrant entrepreneurship may still be nipped in the bud for a number of reasons. Even small outlays of capital may not be forthcoming. Affordable business accommodation for small-scale start-ups may be lacking. Opportunities may also be blocked because the current legislation turns out to be too strict for starting entrepreneurs with only modest educational qualifications. Or established entrepreneurs may exclude newcomers – even if they have the right attitudes and skills (Kehla et al, 1997). This, for instance, has happened in the taxi sector in Amsterdam, where indigenous cab drivers have been able to secure all openings for members of their own social network.

Still, the match between the demand side or the opportunity structure and the supply side consisting of immigrant entrepreneurs is evidently a fact of life in many contemporary advanced urban economies. This applies to the Netherlands as well, where the number of immigrant entrepreneurs in the larger cities has increased considerably in recent years. Below, we will explore the question which processes of *economic* dynamics more specifically underlie this actual expansion of immigrant entrepreneurship.

We will concentrate on the immigrant entrepreneurship in the two largest cities in the Netherlands, Amsterdam and Rotterdam. Both cities are inhabited by large numbers of immigrants (and their direct descendants); by now these make up approximately one third of the entire population of these cities (SCP, 1995: 56). It is likely that the economic transformation processes will manifest themselves most prominently in these cities, as both cities are important nodes – albeit with somewhat different functions – in the *global economy*. Since we are mainly interested in the way opportunities for immigrant entrepreneurship are created, we will concentrate on business start-ups instead of established entrepreneurs.

We will start with a theoretical investigation – on the basis of ideal-typical characteristics of immigrant entrepreneurs – how openings in principle may offer opportunities for setting up a business for those who lack in capital and educational qualifications.

Opportunity Structure and Immigrant Entrepreneurs

The notion of opportunity structure refers to the "demand side" of the "entrepreneur market", just as vacancies for employees refer to the demand side of the labor market (Kloosterman, 2000). The opportunity structure for entrepreneurs is, however, less tangible and less transparent than the demand side of the labor market. Opportunities for businesses are, as a rule, both more diffuse and more obscure. There is, moreover, an essential difference: where potential employees cannot do much more than actively look for and respond to existing vacancies, potential entrepreneurs can, in a way, create openings *themselves* for their own businesses. Contrary to what neo-classic economists tend to think, not all opportunities for businesses are seized, let alone used (Light and Rosenstein, 1995). The "entrepreneur market" is not necessarily in balance and so-called Schumpeterian entrepreneurs can – at least temporarily – enjoy a monopoly by introducing new products or new ways of production, distribution or marketing (Kirzner, 1997). It is also possible that entrepreneurs first set up shop in a sector with relatively low profitability in comparison to other economic activities because they may imitate, for instance, successful relatives or friends or they are funneled towards particular activities by specific social processes (such as discrimination or exclusion from other activities). They may even stick to these unpromising sectors not willing to change tack or not able to switch to new activities lacking in relevant resources in terms of capital, qualifications or social networks. In these cases, the number of enterprises in specific sub-markets can reach its saturation point. However, the limits of the number of entrepreneurs in a particular sub-market may be elastic, but cannot be stretched indefinitely. If the average profitability of a business in a certain market drops to a level (far) below the profitability of alternative uses of labor and capital – let alone if enterprises as a rule make direct losses for a longer period – the influx of new entrepreneurs will dwindle. In the Netherlands, the bottom line will probably be more or less equal to the minimum level of social benefits.[5]

Notwithstanding these clear differences between the labor market and the "entrepreneurial market", the opportunity structure for entrepreneurs can be understood as a hierarchy analogous to Thurow's "job ladder" for the labor market (Thurow, 1975). The opportunity structure is then viewed as a ladder of openings for entrepreneurs ranked according to – potential – profitability and attractiveness (Waldinger, 1996). At the bottom of the ladder of the opportunity structure for entrepreneurs, we will find, hence, the least attractive activities in terms of profitability, working hours, working conditions and prospects for upward mobility.

Entrepreneurs who want to start a business depend on particular parts of this opportunity structure, namely those sectors where economies of scale are not decisive. An (individual) entrepreneur starting as, for instance, an aircraft maker or setting-up an oil refinery is anything but a typical newcomer.

In principle, this condition of viability of small-scale activities applies to all new businesses set up by individual entrepreneurs whatever their origins. Immigrant entrepreneurs, however, usually find themselves in a different position from that of their indigenous counterparts. In the first place, many immigrants from non-industrialized countries have relatively low levels of education or have educational qualifications that are not officially recognized in the country of settlement (SCP, 1996). Setting up a small business in which high-tech skills play a key role is, consequently, not an option to most of these immigrant entrepreneurs.[6] Secondly, it is more difficult for immigrant entrepreneurs to obtain capital for starting a business from banks and other financial institutions (see Waldinger et al. 1990; Wolff and Rath, 2000). Unfamiliarity does not easily make for mutual trust in this respect, and this unfamiliarity applies to the would-be entrepreneurs themselves, as well as to their intended economic activities so that creditworthiness is hard to estimate (Dijk et al, 1993).

Immigrant entrepreneurs, therefore, are dependent on highly accessible, small-scale economic activities that do not require high initial outlays of capital nor specialized knowledge. Labor is the main *input* in these activities. How are opportunities generated for these kinds of activities in advanced urban economies?

Post-industrial Activities or a Game of Musical Chairs?

The most obvious trajectory along which openings for immigrant entrepreneurs are created is a transformation of the economic structure in favor of accessible, labor-intensive and low-value added activities. Processes of globalization and more autonomous developments have indeed profoundly transformed urban economies (Sassen, 1991; Kloosterman, 1996a). The increasing competition of countries with low wages has wiped out a large part of the low-value *manufacturing* activities as it is usually cheaper to import low-value products than to make them in advanced economies (cf. Dicken, 1992). Exceptions to this are markets that are very volatile and where, hence, demand is hard to predict (cf.

Raes, 2000). In these cases, manufacturers often choose a location close to their main market in order to remain informed about fluctuations and shifts in demand and to be able to respond quickly (Scott, 1990). Although in many cases, this concerns high-value, innovative activities, which are not easily accessible to most (would-be) immigrant entrepreneurs, low-value manufacturing is also among these activities that require production locations in or near advanced economies. The sweatshops producing highly fashionable clothing and run by immigrants in Los Angeles, New York, Paris, London and Amsterdam are clear examples of this (see Dicken, 1992; Rath, 1995).

The declining employment in manufacturing has been offset in most cities by a substantial growth of service activities such as producer and personal services, but also wholesale, retail, and restaurants and catering. The expansion of the service sector has inevitably changed the opportunity structure for aspiring entrepreneurs. On the one hand, this post-industrial opportunity structure is characterized by high-value added activities, such as all kinds of producer services (for example, finance, real estate and consultancy) or activities oriented towards the higher echelons of the consumer market (for example, *haute cuisine* and *haute couture*). Starting a business in these high-value markets is on the whole only reserved for highly educated entrepreneurs with very special skills and relevant contacts to acquire – usually indigenous – clients. On the other hand, we also find low-value added activities in these post-industrial cities, both in producer services (for instance, janitors, delivery and parcel services, and security) and in consumer services (for instance, housecleaning, catering, child-care, nail-care, clothing-repair, or valet parking).

For New York, Sassen (1991) has linked social polarization to the expansion of immigrant entrepreneurship. The increase of the share at the top of the distribution of earnings generates the contracting-out of particular tasks creating a demand for low-value service activities. This, according to Sassen, has created opportunities accessible for immigrant entrepreneurs in New York. Processes of social polarization have also occurred in the Netherlands and in Amsterdam and Rotterdam in particular (Kloosterman, 1996a). Although it should be observed that, as yet, the contracting out of all kinds of household tasks by higher income groups has not boomed as much as in the United States, these trends are apparent in Dutch cities as well (see Kloosterman, 1996b; 1998).

The second trajectory along which opportunities are created for aspiring entrepreneurs who have only very limited financial resources and relatively poor educational qualifications, does not necessarily presuppose a transformation of the economic structure. If we envisage the opportunity structure for entrepreneurs as a hierarchy or a ladder, the ladder itself remains the same, as opposed to the interrelated processes of polarization and contracting out as described above. Instead of adding rungs to the bottom or widening existing rungs, openings for immigrant entrepreneurs are created thanks to longer established entrepreneurs who are leaving the bottom rungs. They either retire because of their age or they are able to move up to greener pastures. When indigenous entrepreneurs – and often also immigrants who have been longer established – lack interest in setting-up a business in these relatively unattractive markets, openings are created for newcomers who are prepared to do so.

This so-called *vacancy-chain* mechanism has been analyzed thoroughly by Waldinger (1996). For this process, Waldinger has used the metaphor of a game of musical chairs. Every now and then, a chair is vacated (usually the worst as the upwardly mobile move elsewhere). An immigrant who has just arrived from a less-developed country fills this opening having hardly any choice in the country of settlement. In addition, immigrants from less-developed countries often do not have much resentment towards these specific positions being still attached (at least partly) to sets of preferences of their countries of origin. According to Waldinger, there are usually enough candidates for even the most miserable chair. This, hence, necessitates a selection even at the lower end of the market. To a great extent, social networks of immigrant groups help to determine who can eventually fill a particular position and a *path-dependent* process may be initiated. By means of social networks, vacated places are thus generally allocated to co-ethnics. Thus it becomes possible that so-called *niches* are created: specific markets or sectors that are monopolized by particular groups for a shorter or a longer period of time. These allocation processes whereby ethnic niches may be created can take place without a significant transformation of the urban economy. It does, however, depend on a more or less continuous supply of immigrants who are prepared to undertake the least attractive tasks.

Although both points of view are placed vis-à-vis here, they theoretically do not have to exclude each other and may actually even complement each other very well. A strong growth of opportunities for entrepreneurs higher up in the opportunity structure because of rapid changes in the economic structure can undeniably create room at the lower rungs. The ladder may then become an escalator. Even this is not a necessary condition; room can also be created when the established population turns its back on the city to live in suburbia. This move out of the city may also entail an abandoning of the positions at the lower end of the labor market and the "entrepreneur market" (Cross and Waldinger, 1992). A crucial question concerns the relationship between these two, analytically distinct, processes. This relationship between these processes is important for social mobility prospects. An entrepreneur who can start in a (post-industrial) growth market has entirely different prospects of economic success than a colleague who can only start in a stagnating or even diminishing market that is being left by those who can. This, of course, has also implications for the socio-economic integration of immigrant entrepreneurs as this hinges crucially on the opportunities for upward mobility. An escalator will provide sooner and more opportunities to move upward, than a ladder with static rungs. Below, the position of immigrant entrepreneurs in Amsterdam and Rotterdam will be compared.

Starting Entrepreneurs in Amsterdam and Rotterdam

Economic developments and, consequently, opportunity structures are partly determined, filtered and influenced by the institutional framework (see Kloosterman, 1998; Kloosterman et al, 1999). But, also within a national institutional framework, differences in local economic development occur. Cities, for instance, can have widely divergent economic orientations, structures and histories and this will also have an impact on local opportunity structures. Amsterdam and Rotterdam have a quite diverse economic orientation, in which Amsterdam can be regarded as more post-industrial than Rotterdam. In addition, the Amsterdam economy shows a much stronger general growth of total employment than the Rotterdam economy. In Rotterdam, the strong decline of manufacturing is much less compensated by an expansion of the service sector than in Amsterdam, where trade, catering, business and personal services are doing considerably better than in Rotterdam (cf. Kloosterman, 1996a). Opportunities in Amsterdam, in general but particularly in personal services should be greater than in Rotterdam.

With respect to the demographic side of the equation, it can be said that the process of immigration in Amsterdam and Rotterdam has developed (quantitatively) approximately in the same way and the population of both cities includes a more or less equal share of immigrants. The extent of segregation at neighborhood level is, however, considerably higher in Rotterdam than in Amsterdam.[7] In both cities, the Moroccans and the Turks – large groups in both Amsterdam and Rotterdam – seem to fit best the profile of the ideal-typical would-be entrepreneur who opts for self-employment because relatively poor educational qualifications block many other opportunities (SCP, 1996). These two groups, therefore, are expected to have to resort to the bottom rungs of the ladder of the entrepreneurial opportunity structure. Immigrants from the Antilles and Surinam are, on average, better educated so that, in principle, other sectors will be accessible for them as well.

In order to find out in which sectors entrepreneurs are starting a business, we use data on the number of starting entrepreneurs – Dutch *and* immigrant entrepreneurs – in Amsterdam and Rotterdam in 1994. These figures were supplied by the central data bank of the Chambers of Commerce. Starting entrepreneurs are defined as (legal) persons setting up a new business.[8] The data are based on the trade register of the local chambers of commerce where each new entrepreneur has to register. The extent to which this is actually done cannot be answered with certainty. It is obvious that enterprises which are run entirely informally at home – among them hairdressers, spare-time handymen, and small-scale catering – will not always register but generally starting businesses will do so, if only because registration will offer the entrepreneur particular advantages.[9] The countries of origin have been classified into eight categories: the Netherlands, Surinam, Morocco, Turkey, the Antilles, Cape Verde Islands, Ghana and "Others" (e.g. immigrants from other EU countries, the USA, Canada and Japan). The entrepreneurs about whom this information was lacking form a remaining category.

Although some starting enterprises will develop into successful businesses, and others will give up after a short time, the number of new businesses – regardless of success or failure – is an important indication of "entrepreneurship", or the "entrepreneurial inclination" in a particular society. The sectors where businesses are started can then be seen as an indication of where this inclination manifests itself. We will focus on starting entrepreneurs in the service sector (in the broadest sense of the word) and have left manufacturing and agricultural activities outside of consideration. In all, almost 5,000 new enterprises fall under these sectors.

The "entrepreneurial inclination" in both cities is quite impressive. Table 4.1 shows a survey of the number of business establishments and the number of starting entrepreneurs in the service sector and their share with respect to the total potential work force.[10] In 1994, a total number of 5,000 new businesses were registered in the trade register, more than 3,000 of which were in Amsterdam and almost 2,000 in Rotterdam. Although Amsterdam has a considerably higher percentage of self-employed people than Rotterdam (13 and 5 percent of the potential work force respectively), the percentages of starting entrepreneurs are not very far apart (0.6 and 0.4 percent respectively).

As shown by table 4.2, immigrants constitute about a quarter of all starting entrepreneurs. This strong representation shows a strong entrepreneurial drive among immigrants, which stands out in marked contrast to their unemployment figure, nearly two and a half times as high as that of indigenous Dutch in the four largest cities (CBS, 1996b). It seem likely, therefore, that the relatively poor prospects in the labor market for immigrants is an important motive for setting up their own business (SCP, 1996).

Although the economic orientation of Amsterdam and of Rotterdam differ considerably, the proportion of starting immigrant entrepreneurs in the various sectors, as well as their distribution over these sectors are more or less equal. The exception is catering, where the absolute number of starting entrepreneurs in Amsterdam is higher than in Rotterdam. In both cities, immigrants often choose retail and wholesale trade and catering (cf. Choenni, 1997; Rath and Kloosterman, 2000). Roughly one in every three starting entrepreneurs in the retail and wholesale trade is an immigrant, while their part in the Amsterdam catering industry amounts to 27 percent and no less than 41 percent in the Rotterdam catering industry. In producer services – the sector numbering the most indigenous starting entrepreneurs – only 10 to 12 percent of the starting businessmen is an immigrant. In none of the examined sectors does the number of immigrant starting entrepreneurs exceed those of the indigenous population.

Immigrant Entrepreneurs and the Opportunity Structure

Immigrants do not form a homogeneous group, but differ considerably with respect to migration history, socio-cultural background, educational qualifications, and socio-economic situation. It is likely, therefore, that groups of immigrants will differ with respect to their shares and their distribution among the opportunity structure for entrepreneurs. Tables 4.3a and 4.3b show the distribution of starting entrepreneurs, classified according to country of origin, over the various sectors.

In terms of absolute numbers, Surinamese immigrants and immigrants from the remaining category (particularly in Amsterdam) are the most important groups among the starters in both cities. Turks and Moroccans show considerably less "entrepreneurial inclination". In Rotterdam, Moroccans clearly lag behind Turks, but in Amsterdam the numbers are not very far apart. Both in Amsterdam and in Rotterdam, Turkish and Moroccan starting entrepreneurs are strongly oriented towards trade, both wholesale and retail. If we add catering, we can see that in both cities four out of five Turkish and Moroccan starting entrepreneurs resort to these consumer services. Businesses in consumer services can usually be set-up without great outlays of capital and the strong presence of Turks and Moroccans confirms our predictions (cf. SCP, 1996). The on average higher educated Surinamese are less prone to set up shop in consumer services, instead they opt more for producer services. If we overlook small differences with respect to the large number of Cape Verdeans in Rotterdam and Ghanaians in Amsterdam, both cities show great similarities as regards the ethnic labor division of starting immigrant entrepreneurs. On the face of it, this causes some surprise, since the opportunity structures in Amsterdam and Rotterdam are in some respects quite different.

The level of aggregation when using five sectors is, however, relatively coarse. A more detailed breakdown of the population of starting entrepreneurs is needed. We will, therefore, examine whether these similarities still exist if we use a lower level of aggregation, namely that of subsectors. In total, the Amsterdam and Rotterdam entrepreneurs starting in the services sector are divided into no less than 270 subsectors, ranging from wholesale trade in ice creams and sweets to handymen. For both cities, we have compiled a list ranking the subsectors to the number of starting immigrant entrepreneurs. In table 4.4, we present the ten "most popular" subsectors.[11]

Both in Amsterdam and in Rotterdam, this top ten includes the clothing sector, wholesale in general, grocery sector, snack bars and some forms of service such as driving schools and trade promotion. In order to find out whether the trend of these (sub)sectors is actually towards expansion or whether they represent stagnating or declining markets offering opportunities to immigrant entrepreneurs through *vacancy chains*, we take a look at the development of the numbers of enterprises for each city during the period between 1989 and 1994.[12] In order to see the trends in numbers of enterprises in their proper perspective, the development of the population in Amsterdam and Rotterdam during the period between 1989 and 1994 should be taken into account. In this period, Amsterdam had a population increase of 4.1 percent and Rotterdam 3.7 percent (CBS, 1990; 1991; 1995; 1996).

Taking stock of these trends, we observe the following. In Amsterdam, immigrants appear to start more often in expanding markets. Only two (sub)sectors out of the top ten of starting immigrant entrepreneurs displayed a negative development, namely cleaning agencies and driving schools, while two other sectors are regarded as more or less stagnating (namely, clothing wholesale and snack bars). In Rotterdam, the picture is somewhat different. Three sectors of the top ten show a negative development (retail in vegetables, bars and snack bars), while two others are stagnating (clothing retail and driving schools). The number of strongly growing sectors in Rotterdam is correspondingly lower. Only business consultancy, wholesale in general and dinner delivery show a strong increase.

All in all, in both cities we find starting immigrant entrepreneurs in both expansive and declining or stagnating (sub)sectors.[13] For a large part, these (sub)sectors in the two cities overlap, but the developments in one and the same (sub)sector are sometimes contrary to the developments in the other city. This seems to suggest that the choices of many immigrant entrepreneurs with respect to what kind of business they should start are rather limited and not very much determined by the specific structure of the urban economy. Lacking human and financial capital strongly funnels them towards the same kind of economic activities at the lower end of the distribution. The supply characteristics of the entrepreneurs appear to be more important than the dynamics behind the opportunity structure; they set up shop in retail, restaurants etc, regardless of the structural trends in these sectors. The urban economy does have an impact, however, on the potential success of these businesses: in Amsterdam, the same kinds of businesses are part of an expanding sector, whereas in Rotterdam they are part of a declining one. The Sassen-Waldinger controversy, therefore, hinges not so much on the immigrant entrepreneurs themselves as on the specific economic trajectories of the cities involved.

Urban Economy and Social Mobility

Immigrants prominently take part in setting-up businesses in Amsterdam and Rotterdam although the number of self-employed entrepreneurs in the total work force in Amsterdam is considerably higher than in Rotterdam. Both the percentage of entrepreneurs starting in the service sector and its proportion of immigrants are about the same for the two cities: of the five distinguished sectors taken together, approximately a quarter of the starting entrepreneurs is of foreign origin. This strong representation is another indication of the present rise of immigrant entrepreneurship in the Netherlands and shows that immigrants are not just passive actors waiting for vacant jobs but are actively creating their own employment.

Of these immigrant entrepreneurs, Turks and Moroccans are mainly oriented towards retail and catering, and to a little lesser extent towards personal services as well, while Ghanaians and Cape Verdeans and those from "other countries" are more oriented towards wholesale. Indigenous Dutch and Surinamese are much more oriented towards the structurally expanding producer services. At this level of aggregation, the patterns in Amsterdam and Rotterdam are largely similar.

At a lower level of aggregation, namely that of (sub)sectors, there is also a large overlap. Immigrant entrepreneurs are to be found prominently in (sub)sectors such as wholesale of food and clothing, snack bars and services like schools of motoring and trade promotion in both Amsterdam and Rotterdam. In many cases, these activities are indeed easily accessible in terms of required capital or educational qualifications. Although the (sub)sectors in both cities largely overlap, the developments in those (sub)sectors appear to be different in each city. It is here, that we do see some intriguing differences. The top ten of (sub)sectors where immigrant entrepreneurs set up their businesses in Amsterdam are more liable to be expanding sectors than those in Rotterdam.

Immigrant entrepreneurs, therefore, seem to be dependent on the same kind of openings for starting a business. The specific dynamics of the urban economy, however, determines, to a large extent if these businesses are more part of dead-end street vacancy chain processes as proposed by Waldinger, or more part of expansion of low-value added activities as described by Sassen. Given the specific processes of funneling of immigrants towards specific activities, the development of the local economy does not seem to determine the labor division of immigrants, but rather the chances of success as entrepreneur in those specific sectors.

Although it is still premature to make more definite statements on entrepreneurship *as an avenue of social mobility*, our findings show that independent entrepreneurship in advanced urban economies does not necessarily have to be a dead-end street for immigrants. Many immigrant entrepreneurs are found in saturated markets where cut-throat competition prevails, but this is certainly not the whole story. In Amsterdam, and to a lesser extent in Rotterdam, we can see that immigrants can certainly be found in highly expansive sectors as well, with much better prospects of success. A permanent marginal existence as *lumpen bourgeois* is evidently not the only possible outcome of immigrant entrepreneurship in the Netherlands. This implies that the scope for immigrant entrepreneurs becoming agents of regeneration of deprived neighborhoods could be considerable on the condition that such businesses will – at least for significant period – stay put. Thus, not only neighborhoods may change but also the perception of the minorities by the dominant groups and even their self-perception shifting the connotation of "immigrants" and "minorities" along the way.

Notes

1   This research project is part of Working on the Fringes; Immigrant Businesses, Economic Integration and Informal Practices, a thematic European network for exchange of knowledge and experiences. This international network, co-ordinated by Jan Rath and Robert Kloosterman and funded by the European Commission under the Fourth Framework, involves both international comparison and collaboration with regard to research on immigrant entrepreneurs in Austria, Britain, France, Germany, Israel, Italy and The Netherlands. This research is also part of Immigrant Self-Employment, Mixed Embeddedness, and the Multicultural City research program funded by the NWO under the MPS heading. We would like to express our thanks to Jan Rath for his invaluable comments.

2   Sassen, 1991; Ma Mung, 1992; Barrett et al, 1996; Waldinger, 1996; Waldinger and Bozorgmehr, 1996 and Simon, 1997.

3   Now that the increase of the number of starting entrepreneurs continues in times of a strong, prolonged boom, starting a business cannot be regarded anymore as a defensive choice only. During the period between 1992 and 1996, the number of self-employed people increased from 627,000 to 728,000, or from 10.6 percent of the total work force to 11.8 percent (CBS, 1998).

4   This holds particularly true for the Netherlands. According to Waldinger et al (1990: 152), starting a business in the Netherlands in the 1970s and 1980s was rather difficult because of the "maze of regulations that govern business activities in the Netherlands". More recently, however, deregulation has made a big dent in this maze. Starting a new business in the Netherlands is, according to the World Economic Forum, only slightly less difficult than in the United States and almost as easy as in Britain (*The Economist*, 1999: 143).

5   In cases it may even drop below this line as entrepreneurs may prefer independence as, for instance, a shop-owner to either living on welfare or being employed as a low-skilled laborer.

6   Times can change, though, as is shown by the recent wave of establishments of software companies by Chinese and Indian entrepreneurs in Silicon Valley, California (*The Economist*, 1997). As yet, these entrepreneurs with their extremely sophisticated intellectual capital are anything but representative of the great majority of immigrant entrepreneurs from less-developed countries.

7   The segregation index of minorities at neighborhood level in Amsterdam amounted to 31.2 in the mid-nineties and 43.3 in Rotterdam (SCP, 1995, p. 64).

8   Immigrant entrepreneurs are considered to be persons with another nationality than Dutch. For entrepreneurs of the Antilles, Surinam, the Cape Verde Islands and Ghana, the country of birth has been considered as well, in addition to nationality. The category

"other countries" includes both entrepreneurs from industrialized countries and people from non-industrialized countries. This category is considerably larger in Amsterdam than in Rotterdam. From various analyses, it appears that persons about whom no information as to nationality and/or country of birth was available did not score spectacularly, they are quite close to the average. This means that systematic bias because of this lacking information is unlikely to occur in the data. Some entrepreneurs have an additional paid job.

[9] In spite of the inevitable drawbacks of such registrations, the data of the trade register are still the most suitable for obtaining an overview of new immigrant enterprises in the Netherlands (see also Setzpfand et al, 1993; SCP, 1996; Choenni, 1997). We should take into account that entrepreneurs without the proper documents tend to register their business as wholesale, rather than retail and that it will take some time before such entrepreneurs are actually starting.

[10] For an analysis of the spatial patterns of starting immigrant entrepreneurs, see Kloosterman and Van der Leun, 1999.

[11] In Amsterdam, the two sectors taking tenth place *ex aequo* are both presented.

[12] A further breaking down into immigrant categories would have yielded too small numbers.

[13] One could say that the increase of businesses is caused by the influx of immigrant entrepreneurs. This holds true to a certain extent only. Firstly, we have inferred above that a given opportunity structure eventually limits the number of entrepreneurs in a specific segment of the market as firms in the long run have to be profitable no matter how slightly. Secondly, this would mean that all top ten sectors would have to show an increase, which is certainly not the case.

References

Barrett, G. A., Jones, T. P. and McEvoy, D. (1996), "Ethnic minority business: theoretical discourse in Britain and North America", *Urban Studies*, 33(4/5), pp. 783–809.

Castles, S. and Miller, M. J. (1993), The Age of migration; International Movements in the *Modern World*, Hampshire: Macmillan.

CBS (1990), *Statistisch Jaarboek 1990*, Heerlen/Voorburg: Centraal Bureau voor de Statistiek.

CBS (1991), *Statistisch Jaarboek 1991*, Heerlen/Voorburg: Centraal Bureau voor de Statistiek.

CBS (1995), *Statistisch Jaarboek 1995*, Heerlen/Voorburg: Centraal Bureau voor de Statistiek.

CBS (1996a), *Statistisch Jaarboek 1996*, Heerlen/Voorburg: Centraal Bureau voor de Statistiek.

CBS (1996b), *Allochtonen in Nederland 1996*, Heerlen/Voorburg: Centraal Bureau voor de Statistiek.

CBS (1998), *Enquête Beroepsbevolking (EBB)*, Heerlen/Voorburg: Centraal Bureau van de Statistiek.

Choenni, A. (1997), *Veelkleurig assortiment; Allochtoon ondernemerschap in Amsterdam als incorporatietraject 1965–1995*, Amsterdam: Het Spinhuis.

Cross, M. and Waldinger, R. (1992), "Migrants and Minorities and the Ethnic Division of Labor", in Fainstein, S. S., Gordon, I. And Harloe, M. (eds), *Divided Cities: London and New York in the Contemporary World*, Oxford: Blackwell.

Daniels, P. W. (1993), *Service Industries in the World Economy*, Oxford: Basil Blackwell.

Dicken, P. (1992), *Global Shift; The Internationalization of Economic Activity* (2nd Edition), London: Paul Chapman Publishing Ltd.

Dijk, S., Van Genus, R. C. and Noordemeer, H. (1993), *Allochtone ondernemers en het bijstandsbesluit zelfstandigen*, Den Haag: Vuga.

Economist, The (1999), "Financial indicators; new businesses", 16th October, p. 143.

Engbersen, G. and Van der Leun, J. P. (1998), "Illegality and criminality: the differential opportunity structure of undocumented immigrants", in Koser, K. and Lutz, H. (eds), *The New Migration in Europe*, Hampshire: Macmillan pp. 199–223.

Engels, W., Linssen, P. and Setzpfand, R. (1993), *Onderzoek vraagzijde. Deelrapport I bij het onderzoek naar de effecten van het beleid inzake het ondernemerschap van allochtonen*, Utrecht: Coopers and Lybrand Management Consultants.

Holliffield, J.F. (1992), *Immigrants, Markets and States; The Political Economy of Postwar Europe*, Cambridge MA/London: Harvard University Press.

Kehla, J., Engbersen, G. and Snel, E. (1997), *"Pier 80" Een onderzoek naar informaliteit op de markt*, Den Hag: Vuga.

Kirzner, I. M. (1997), "Entrepreneurial discovery and the competitve market process; an Austrian approach", *Journal of Economic Literature*, 35(1), pp. 60–85.

Kloosterman, R. C. (1996a), "Double Dutch: trends of polarization in Amsterdam and Rotterdam after 1980", *Regional Studies*, October 1996, 30(5), pp. 367–376.

Kloosterman, R. C. (1996b), "Mixed experiences; post-industrial transition and ethnic minorities on the Amsterdam labor market", *New Community*, December 1996, 22(4), pp. 637–653.

Kloosterman, R. C. (2000), "Immigrant entrepreneurship and the institutional context. A theoretical exploration", in Rath, J. (ed.), *Immigrant Businesses. The Economic, Politico-Institutional and Social Environment*, Hampshire: Macmillan, pp. 135–160.

Kloosterman, R. C. and Van der Leun, J. P. (1999), "Just for starters: Commercial gentrification by immigrant entrepreneurs in Amsterdam and Rotterdam Neighborhoods", *Housing Studies*, 14(5), pp. 659–676.

Kloosterman, R.C., Van der Leun, J. P. and Rath, J. (1998), "Across the border: Economic opportunities, social capital and informal business activities of immigrants", *Journal of Ethnic and Migration Studies*, 24(2), pp. 249–268.

Kloosterman, R. C., Van der Leun, J. P. and Rath, J. (1999), "Mixed embeddedness, migrant entrepreneurship and informal economic activities", *International Journal of Urban and Regional Research*, 23(2), pp. 253–267.

Light, I. And Rosenstein, C. (1995), *Race, Ethnicity and Entrepreneurship in Urban America*, New York: Aldine de Gruyter.

Ma Mung, E. (1992), "L'expansion du commerce ethnique: Asiathiques et Mahrébins dans la régions parisienne", *Revue européenne des migrations internationales*, 8(1), pp. 39–59.

McEvoy, D., Barrett, G. A. and Jones, T. P. (1998), "Market potential as decisive influence on the performance of ethnic minority businesses", in Rath, J. (ed.), *Immigrant Businesses on the Urban Economic Fringe. A Case for Interdisciplinary Analysis*, Hampshire: Macmillan.

OECD (1993), *Employment Outlook 1993*, Paris: OECD.

Raes, S. (2000), *Migrating Enterprise and Migrant Entrpreneurs; A Study into the Role of Fashion and Migration in the Changes in the International Division of Labor in the Clothing Sector from a Dutch Perspective*, Amsterdam: Het Spinhui.

Rath, J. (1995), 'Beunhazen van buiten. Die informele economie als van sociale intergratie', in Engbersen, G. and Gabriëls, R. (eds), *Sferen van integratie; Jaarboek Beleid en Maatschappij*, Meppel: Boom.

Rath, J. (1999, forthcoming), "A game of ethnic musical chairs? Immigrant businesses and the elleged formation and succession of niches in the Amsterdam economy", in Body-Gendrot, S. and Martiniello, M. (eds), *Minorities in European Cities. The Dynamics of Social Integration and Social Exclusion at the Neighbourhood level*, Hampshire: Macmillan Press.

Rath, J. and Kloosterman, R. (2000, forthcoming), "'Outsiders business' A critical review of research on immigrant entrepreneurship", *International Migration Review*.

Rekers, A.M. (1993), "A tale of two cities. A comparison of Turkish enterprises in Amsterdam and Rotterdam", in Crommentuyn-Ondaatjie, D. (ed.), *Nethur School Proceedings 1992*, pp. 45-66, Utrecht: Nethur.

Sassen, S. (1991), *The Global City: New York, London, Tokyo*. Princeton (NJ): Princeton University Press.

Scott, A.J. (1990), *Metropolis; From the Division of Labor to the Urban Form*, Berkeley/Los Angeles/London: University of California Press.

SCP (1995), *Rapportage minderheden 1995; Concentratie en segregatie*, Rijswijk: Sociaal Cultureel Planbureau.

SCP (1996), *Rapportage minderheden 1996; Bevolking, arbeid, onderwijs, huisvesting*, Rijswijk: Sociaal Cultureel Planbureau.

SCP (1998), *Rapportage minderheden 1996; Bevolking, arbeid, onderwijs, huisvesting*, Rijswijk: Sociaal Cultureel Planbureau.

Thurow, L. (1975), *Generating Inequality; Mechanisms of Distribution in the U.S. Economy*, New York: Basic Books.

Waldinger, R. (1996), Still the Promised City? *African-Americans and New Immigrants in Postindustrial New York*, Cambridge (Ma.)/London: Harvard University Press.

Waldinger, R. and Bozorgmehr, M. (1996), *Ethnic Los Angeles*, New York: Russell Sage Foundation.

Waldinger, R. and Lapp, M. (1993) "Back to the sweatshop or ahead to the informal sector?", in *International Journal of Urban and Regional Research*, 17(1):6-29.

Waldinger, R., Aldrich, R., Ward, R. and Associates (1990), *Ethnic Entrepreneurs. Immigrant Business in Industrial Societies*, Newbury Park: Sage.

Wolff, R. and Rath, J. (2000), *Centen tellen: Een inventariserende en verkennende studie naar de financiering van immigrantenondernemingen*, Amsterdam: Het Spinhuis.

Zanden, J.L. van (1997), Een klein land in die 20e eeuw; Economische geschiedenis van Nederland 1914-1995, Utrecht: Het Spectru.

Table 4.1   Number of entrepreneurs out of the work force and the number of starting entrepreneurs in the services sector

	Establishments excl. agriculture	Potential work force	Number of entrepreneurs out of the work force	Total number of starting entrepreneurs	Percentage of starting entrepreneurs out of the work force
Amsterdam	67,733	515,800	13	3,158	0.6
Rotterdam	21,419	401,105	5	1,792	0.4

Source: Calculated on the basis of data provided by O+S 1994, COS 1994 and NV databank Woerden 1994.

Table 4.2   Starting entrepreneurs classified according to sector, city and percentage of immigrant entrepreneurs, 1994

Sector	Amsterdam		Rotterdam	
	Number of starting entrepreneurs	Number and percentage of immigrants (%)	Number of starting entrepreneurs	Number and percentage of immigrants (%)
Wholesale	814	286 (35)	524	159 (30)
Retail	676	193 (28)	332	100 (30)
Producer services	1033	126 (12)	544	54 (10)
Catering	209	57 (27)	165	67 (41)
Personal services	426	82 (19)	227	67 (30)
Total	3158	744 (24)	1792	447 (25)

Source: Calculated on the basis of data provided by NV databank Woerden.

Table 4.3a Amsterdam starting entrepreneurs in the services sector, classified according to country of origin and branch (absolute numbers), 1994

	NI	Tur	Mor	Ant	Sur	CV	Gha	Oth.	Unclass	Tot.
Wholesale	382	14	12	8	44	0	23	185	146	814
Retail	367	21	36	6	33	0	4	93	116	676
Producer services	788	3	2	2	33	0	4	82	119	1033
Catering	113	8	5	1	17	0	2	24	39	209
Pers. services	283	10	7	2	24	0	4	35	61	426
Total	1933	56	62	19	151	0	37	419	481	3158

Source: Calculated on the basis of data provided by NV databank Woerden.

Table 4.3b Rotterdam starting entrepreneurs in the services sector, classified according to country of origin and branch (absolute numbers), 1994

	NI	Tur	Mor	Ant	Sur	CV	Gha	Oth.	Unclass	Tot.
Wholesale	323	29	8	5	35	5	3	74	42	524
Retail	209	30	17	4	31	1	1	16	23	332
Producer services	459	2	2	3	28	0	0	19	31	544
Catering	91	15	13	4	19	0	0	16	7	165
Pers. services	153	14	5	3	33	3	1	8	7	227
Total	1235	90	45	19	146	9	5	133	110	1792

Source: Calculated on the basis of data provided by NV databank Woerden.

Table 4.4   Top ten sectors with the most immigrant starting entrepreneurs in 1994, with the percentage change of the number of entrepreneurs during the period between 1989 and 1994 in Amsterdam and Rotterdam

	Amsterdam	Sectoral trend	Rotterdam	Sectoral trend
1.	clothing retail	17	general wholesale	69
2.	clothing wholesale	5	driving schools	9
3.	general wholesale	87	clothing wholesale	16
4.	groceries wholesale	50	snack bars	-8
5.	household goods retail	23	bars	-10
6.	handymen	73	clothing retail	2
7.	trade promotion	27	catering	30
8.	cleaning agencies	-8	cleaning agencies	12
9.	driving schools	-5	vegetables retail	-29
10.	snack bars	8	business consultancy	86
11.	catering	63		

Source: Top ten and trends calculated on the basis of data provided by NV Databank Woerden.

Chapter 5

Historic Preservation in New Orleans' French Quarter: Tolerance and Unresolved Racial Tensions

JOHN FOLEY AND MICKEY LAURIA

This chapter deals with the controversies which surround a minority's efforts to confront development pressures in an historic area – the French Quarter (Vieux Carré) in New Orleans. Their efforts seek, as well as preservation, the maintenance of a diverse residential base. They see this as challenged both by the standardizing tendencies generated by tourist development and through the displacement caused by increasing property values in general. However, the aims of this minority have to be seen in the context of the values of other minorities even when such generally accepted issues as historic preservation are involved. Controversies remain as to the preference that should be given to preservation over development.

Historic Preservation, Urban Revitalization: Controversies

Historic preservation has not always been an accepted theme (Gamson, 1988, p. 220), around which groups organize. Now it is a respectable cause, the defense of which has entered the mainstream (Lofland, 1996, p. 9). Opposition to such causes is usually muted and generates little openly expressed resistance. However, preservation activism can still be interpreted as oppositional to traditional concepts of development (Caufield, 1994, p. xiv; Thomas, 1994, p. 69) as typified in the theoretical constructions of growth machines (Logan and Molotch, 1987) and developmental regimes (Stone, 1993). For this reason, groups defending preservation appear in opposition to economic and political interests promoting their vision of economic development. In New Orleans' French Quarter, groups and individuals defending preservation, despite their affluence and dominantly white membership, see themselves as an under-represented minority confronting a dominant discourse favoring development.

At the same time, the oppositional character of this "minority" has to be seen in the context of their defense of certain forms of capital. This occurs without those involved being fully aware that they are seeking to maximize their specific profits (Bourdieu, 1993, p. 76). As such, there is a defense of determined values about the way the world should be, always in some measure reflecting the social class origins (Barthel, 1996, p. 3) of activists. Thus, establishment of historic areas' identity is "likely to reflect the perceptions and interests of the powerful and well established"

(Thomas, 1994, p. 71). Given these circumstances, the defense of preservation has to be seen in the context of the values of other "minorities".

In New Orleans the black population is a minority in the French Quarter but is a majority in Orleans Parish[1] as a whole. The racial composition of the City Council reflects this dominance as all but one of the representatives' is/are black, as is the Mayor. These groups represent a continuum in the priority given to tourist development. There is no significant difference between policies of black and white political administrations as far as tourist development is concerned. The present administration does, however, promote a discourse supportive of policies, which generate opportunities for the less affluent black community and in turn receives its ample popular support. Such policies can be seen in the context of a city increasingly dependent on tourism as a source of economic growth (Brooks and Young, 1993, p. 268) and employment opportunities.

The French Quarter is an essential part of this promotion of New Orleans as a tourist and convention center. Along with the Warehouse District (a former industrial area being converted to residential, cultural and tourist activities) and the CBD, it is the site of major investments. This includes most recently the construction of an aquarium, extension of the convention center, a new stadium alongside the Superdome, and in November 1999 the opening of the huge Harrah's casino, only one block from the French Quarter. Such development has placed considerable pressures on the Quarter because of the scale of adjacent development (Brooks and Young, 1993, p. 254), the rapid increase in the number of tourists, and the threat to its diverse residential and commercial base.

The development discourse legitimates decisions that do not give priority to preservation interests. Decisions appearing to promote tourist (and other forms of economic development) frequently alienate the white minority favoring preservation, particularly of the historic residential areas in the city. The idea that the mainly white groups aligned to the preservation ideal is treated like a minority – although obviously not an unprivileged one – is reinforced. They portray themselves as facing a black majority with different values in relation to the relative importance of preservation and development. Conflicts remain unresolved but cannot be seen in dichotomous terms of a classic struggle between classes or between an underprivileged minority confronting an all-powerful majority (or a majority facing an all-powerful minority).

To continue with the racial aspects, historic preservation is often associated with processes of urban renovation, described variously as gentrification, the back-to the-city movement, or urban revitalization. Frequently, the process is portrayed as causing expulsion of poor and minority people. No sooner was this type of urban revitalization identified than concern was expressed for the way existing residents were progressively displaced by newcomers who had the economic resources necessary to buy, restore and maintain historical areas. Black residents were often the victims (Spain, 1980, p. 39), reducing diversity and creating new forms of segregation (Laska and Spain, 1980b, p. 132 – in New Orleans; Tournier, 1980, p. 174 – in Charleston; Gale, 1980, p. 96 – in Washington).

In New Orleans' French Quarter census figures for the last fifty years show clearly a process of displacement. For instance, the proportion of black population in the Quarter declined from almost 20 percent in 1940, to just under five percent in

1990. In absolute terms these figures show an even more dramatic decrease. By 1990 only 190 black residents lived there while in 1940 there were 2,179. Such a decline is part of the substantial loss of residential population as a whole in the Quarter. Population has decreased from around eleven thousand to two thousand in the same period. So black residents appear disproportionately affected by the changes that have occurred in the Quarter. As a point of reference, in this period, the proportion of black population in the Orleans Parish has increased from around 30 percent to more than 60 percent of a total population of almost half a million (CUPA, 1992, A-10, see also Wilkenson, 1985). Similarly, Quarter residents now come from much higher income groups (CUPA, 1992, A-15).

Nonetheless, the regeneration process is interpreted as benefiting the resurgence of US inner cities. Shirley Laska and Daphne Spain (1980a, p. xiii) point specifically to the increased tax base and the greater trust in the viability of inner cities. Similarly, when these are areas of architectural and historic significance, the effect of securing their preservation is an additional positive argument. Such aims have been achieved in the French Quarter. Thus, the substantial decrease in racial diversity has to be seen in the context of the Quarter's change from the "giant slum" of the 1920s and 30s. Today, it is a preserved mixed development area attractive to upper income residents and a unique historic area magnet to millions of tourists.[2]

The benefits and costs of such processes of revitalization can be seen, then, from different points of view. Whatever interpretation is given, what is clear is the role of French Quarter preservationists in the conservation of this unique area. As one resident puts it, "some of these people really probably saved the place... It would have been long gone if these preservationists in the 1920s and 1930s hadn't turned things round".[3] Walter Gallas (1996) substantiates this positive interpretation of the vital role of neighborhood activists in the preservation of the Quarter.[4]

Thus, the implicit negative interpretation of gentrification is modified by the need to balance positive and negative aspects of the preservation of historic areas and to consider the important role played by preservationists. Jon Caufield (1994), following a social movement interpretation of neighborhood revitalization,[5] suggests the need to place middle class resettlement in historical context that takes into account the opposition to certain groups' growth policies. This is one justification for the social movement interpretation (Klandermans and Tarrow, 1988). For Caufield, for instance, "middle-class settlement of older inner-city neighborhoods has, in part, constituted an urban social movement". For although incrusted in the logic of property capital it represents "an immanent critique of key facets of contemporary city-building" (p. xiv). Such findings are supported in the French Quarter where it is also sustained that alternative lifestyle development, and the protection of the safe space for certain minorities, particularly the gay community, justifies this social movement interpretation (see Foley, 1999).

Despite these positive interpretations, the question of race and the values involved in preservation movements cannot be totally evaded. The following discussion will point to the potentially conflicting visions that show the need for us all to look at structures of meaning. As Patsy Healey (1997, p. 93) points out, we should look at the "'deep structures' of power embedded in our ways of thinking and organizing". Not doing so "could have the effect of unwittingly reinforcing the power relations and driving forces that are constraining the invention of new practice". So when diversity marks a population, it becomes necessary to consider how different racial, ethnic or lifestyle groups perceive historic preservation so that what is being preserved in not just a reminder of a history of prejudice and segregation. It points, also, to the need to consider a more ample context for the practice of historic preservation.

The French Quarter: a Celebration of Diversity

The French Quarter consists of a 97–block area hugged by the Mississippi's curving crescent. In an intensely mixed-use area, of diverse character, tourist, commercial and residential activities exist in close proximity. Such diversity is embedded in a physical structure, equally varied, fashioned by builders during more than two hundred years.[6] Despite this, residential activity still dominates.[7] Alongside the diversions, especially associated with world-renowned Bourbon Street, people live permanently. Houses and a large variety of commerce mix in a fascinating jumble. One part of a street will house sex and tee-shirt shops while nearby are antique shops, art galleries and designer clothing stores for the most refined tastes. Neighbors and visitors can appreciate this mixture of spaces and activities, allowing pedestrian access to most of the needed daily services as well as to extraordinary diversions satisfying disparate passions. Meanwhile, behind the commerce people live around their patios. Above, balconies give constant access to a passing world. The scale is modest and domestic. A variety of architectural styles, an abundance of detail, buildings, patios and public spaces, coalesce into an appealing whole.

It becomes easy to see why those connected to the Quarter credit the physical structure as being exceptionally important in making the area what it is today. Within this shell successive generations of activities and people have shaped a fascinating place. Space, people and activities join inextricably and form a past that cannot be separated from a consciousness of the present. Quarterites are imbued with a history that comfortingly places them in relation to the ordinary and extraordinary personalities who have been living there. So neighbors are conscious of this connection between the physical structure, activities and the existence of diverse lifestyles. They understand that preservation of the historic district demands attention to all these aspects and insist that preservation be not just "bricks and mortar" but include as well the maintenance of population diversity. Diversity of population in terms of occupational groups, races, ethnicities and lifestyles is understood by many to be one of the principal features which makes the Quarter unique. As one resident, Andrew, expresses it, "people do what ever they do as long as it doesn't affect the public well-being if you will". Tolerance of lifestyles is part of a concern for the maintenance of diverse occupational groups. He adds, "it's really the only place that I have ever found where, you know, bank presidents live next to bar tenders". For Danny such tolerance includes racial. "It is just a place where everyone is welcome, regardless of who they are, what their race, and what their economic status".

This mixture of activities and people, brings certain problems but instills the Quarter with a very special character. It makes it, not just a static museum-like place, but as one resident calls it, "a living, breathing, historical district". She points out, "We don't open up at nine o'clock in the morning and put on costumes and then close at five o'clock in the afternoon, and go home. We live here". With these sentiments she represents the view that this is "home" and it is a home that will be defended. It is a home with permanent occupiers, where people live who are constantly active in the neighborhood. It contrasts with "most historical districts in other cities" where, "it's a couple of store front buildings… maybe they even got a couple of people working during the day in a blacksmith shop, or making baskets… mostly for show". Resulting from such sentiments, the battle cry for Quarterites becomes: "this is a living, breathing neighborhood". So their aim is to oppose what Michael Thomas (1994, p. 65) calls a process where certain cities becoming museums, functioning only as tourist attractions. Neighbors value their role of defenders of this vitally active neighborhood and, merely by living there, can be considered as upholding its residential tradition.

The Neighborhood Organizations

It is in this context that the struggle for preservation of the French Quarter has been organized. For this the Vieux Carré Property Owners Association was formally consolidated in 1938 (Gallas, 1996), although activism can be traced back earlier. Lyle Saxon (1995, p. 273), writer and chronicler of the Quarter, illustrates this and the fact that historic preservation was considered, at best, a pursuit of cranks. He writes of an early preservation effort in which he was involved. "At the corner of Royal and St. Louis streets… the old St. Louis Hotel… was torn down in 1917… it should have been preserved, for it was a beautiful building. I remember how a group of us tried to save it from destruction and how we were laughed at for our pains."

Present day activists assign much credit for the consolidation of a preservation group to such individuals.[8] They legitimized preservation as a valid concern at a time when it was not a commonly accepted value. As one put it, these were the "people who saved the butt of the city by protecting this neighborhood years ago". Similarly, it was pressure from such individuals that persuaded the State to establish institutional oversight in 1925. The first Vieux Carré Commission was consolidated in New Orleans to oversee development in its jurisdiction. They established a new Vieux Carré Commission in 1937 with enabling legislation that allowed it to control modification to existing buildings and new construction.

The early preservation group had been consolidated, under its 1976 name, as the Vieux Carré Property Owners, Residents and Associates. It has been a leading voluntary organization that represents preservation interests in the French Quarter. It has a membership of about 550, its own office, a halftime official and regularly defends the preservation ideal. Today, a number of other groups have formed concerned with particular issues, for instance Crime Watches, or with specific spatial areas within the Quarter: Jackson Square or the Upper Quarter.[9]

Nonetheless development pressures have continued throughout this century. An early commentator on the French Quarter, Lyle Saxon complained that for the present Agriculture and Fisheries Building on the 400 block of Chartres "a whole square of delightful old houses was destroyed" (Saxon, 1995, p. 271). The redevelopment was a symbol to local boosters who saw it as an example that would encourage development.[10] Not even in the post-war years did development pressures cease. One informant, remembers that around the end of the decade of the 1950s, the city's then mayor, Morrison, still supported his Safety and Permits' director who stated that "they should bulldoze the whole Quarter".

Despite the increasing acceptance of the preservation doctrine, in the 1960s preservation groups fought a major battle against a proposed elevated freeway, that would have passed between the French Quarter and the Mississippi River. Richard Baumbach and William Borah (1981) called it the "Second Battle of New Orleans". It "became more than just a conflict between environmentalists and downtown developers over a freeway: it was a clash of values, a clash in attitudes, a difference in priorities and perspectives about the character of the city" (p. 3). This struggle became a "cause celebre" and drew to it many organizations including those at a national level. When the proposal was withdrawn in 1969, the US Secretary of Transportation justified the decision as the beginning of a new tradition of preserving the nation's heritage (Baumbach and Borah, 1981, p. xiii). Preservationists considered it a landmark victory.

Thus groups and individuals defending preservation have been an essential part of the way the Quarter has been transformed from the "giant slum" of earlier times. What has changed is the concern not just for preservation of the physical structure but for the quality of life issues and the recognition that much of the Quarter's character is derived from its diversity both of population and activities. In recent times, it is through pressures, including street demonstrations, that regulations banning large buses have been introduced. Similarly, activists demanding stop signs to control traffic speeds, have been successful in getting them erected. Now an important preoccupation for these organizations is the decline of permanent population in the Quarter, seen as derived from the uncontrolled proliferation of commercial uses. Confronting these processes becomes more difficult because they become embroiled in legal proceedings that require professional legal representation (as will be seen later when discussing the problems of enforcement of city ordinances).

At the same time, the very success of the historic preservation of the physical structure brings increasing commercial pressures from tourist activities. These pressures represent a threat both to diversity of activities and population. A "delicate balance" is achieved that is innately unstable. Not surprisingly, balance between commercial and residential activities is not easy to maintain and differences exist as to where the fulcrum should be.

Divergent Visions and Unresolved Racial Tensions

The priority given to different worldviews becomes clearer in the analysis of conversations with a group of denizens. The segregated past still affects the perception of the Quarter by the citywide black majority population, and it is not a place where they feel comfortable to live. In the treatment of certain issues, close to the hearts of Quarterites, such as noise and crime, more racial tensions are generated. Even the very discourse of development is given racial connotations.

The Black Community in the Quarter

We have already described the displacement of black population. Residents are aware of this and, because there are few black residents, they tend to "stand out" and be known. Informants speak of their black neighbors with affection, worried they will be displaced, so converting the Quarter into a truly white residential ghetto. Fear exists both of losing diversity and of the political consequences of being seen as a "white elite enclave".

Some residents point to the important role-played by the black population, particularly black Creole artisans, in the construction of the Quarter. Not only were they important as artisans but owned, also, a substantial number of properties. One neighborhood activist speaks of doing research and finding many buildings owned by free persons of color. She found it "was unbelievable. I am talking about a very high percentage, of the whole Quarter, even the elite homes". So for her black people "did have their niche here".

White residents feel no disharmony between black and white people living in the Quarter. This co-existence is illustrated by one homeowner who tells how the presence of black people is not thought of as a threat when he speaks of the clients of a black gay bar: "it's just a loud crowd... I don't mind that in as much as I feel like if people are out on the sidewalk, they're keeping it safe, and it's not a criminal crowd at all". In fact, their presence is felt a symbol of security. However, he does make it clear that it is not his "social crowd". Such a comment intimates that, generally, there still exists a clear division between racial groups in their social activities. The separation between "us" and "them" remains.

A woman who owns property in the Quarter also wants to see the positive side and states that "despite the politicians there is a great deal of goodwill between the races, don't you think?". However, she can note the hostility that does exist toward her: "well I do think there are young blacks who have a lot of hostility and I see this mostly in the grocery store, where they are really are not as nice as they should be". Clearly tensions exist and the established white population may see young black men, who they do not recognize from the neighborhood, as a potential threat. Resentment builds up on both sides. As groups we depend upon our individual characteristics and resent the fact that we too are essentialized as racists at the same time that we essentialize the other. Our defense against racism is our individual tolerance. She goes on: "and I think that sometimes I may be perceived by these people to be someone, because of my age, that I am a racist, which is something I have never been at all. I sent my children to public schools in the Mississippi delta that are primarily black". She adds that, "White people live here by choice and if they hated black people, they wouldn't be here". Such comments show that racism is seen in terms only of outright hatred and not in the way the dominant culture maintains predominance in determining what are acceptable values. She finishes saying "it's sort of absurd for black people to see me as an enemy". Remembering that we have responsibilities both as a class, and as individuals, is difficult.

Another theme is the need to educate black people so that they will come to appreciate the same things as the white population, such as the historic value of preservation of the French Quarter. Education sounds often like the desire to instill values without reflexion on their cultural bias. This is an extreme interpretation but when remarks are extracted from their context they can be interpreted this way. For instance, people speak of "educating black people about their history and the history of the city", or meeting Treme [11] people "over the children musicians and dancers... to give them an education and all, so that they cannot be the band from hell".

Nonetheless, open expressions of racism are rarely observed.[12] Respondents prefer to speak of other people's racism. For some "real antipathy" still exists. One spoke of a neighborhood effort to improve the local McDonough School's garden and those that did not want to participate because "it was for black kids", but she does say that they are in a minority now.

As mentioned, the organized community worries that they could be seen as racist and elitist by a black political establishment, and the public overall. For David, "there are race and class issues for this, because to own is very expensive, as well as the maintenance, so for lower income families it is impossible. Even for renting it is not affordable for such families. This means that at times the French Quarter is seen as a white enclave of elitists who are trying to keep out the riffraff. Some of this is true. It is true that it is elitist".

Such a reputation could be the reason why black people, even those with the resources, do not choose the Quarter as a place to live. Some of this could be the result from the Quarter was once out of bounds for black people as an area of diversion. A black resident tells this story. "I was born in Treme... in that period (I calculate in the late 60s),[13] people who were considered colored, non-whites, weren't actually allowed in the Quarter, because everything was segregated. So even growing up here, the Quarter for me, was not user-friendly until the days of desegregation... I was barred legally from going in there unless I said I was white and not a person of color. For that one reason, the Quarter was always off bounds to me. Even until today there are many African-Americans who see Rampart Street as the Mason-Dixon Line."[14] He points to the existing legacy of this segregation. "You are always going to experience discrimination, even like recently, you know I was in a (French Quarter gay) bar one night, and you see black people coming in at the door and they were asking for three or four IDs so I went over to the manager and said; "you've got the wrong person on the door. Where is this idiot from?" Well he had been trained to do that at another straight bar up the street. And so the manager called him on the side, in front of me, and said, 'we don't do that down here'. And he apologized". So there is an increased awareness that discrimination does exist and on these occasions something is done about it.[15]

The reference to Treme and the existence of a "Mason-Dixon line" can be further examined through its relationship to the Quarter. Consciousness of physical separateness and racial segregation persists. Initially, a Quarter activist points to the fact that relations between the neighborhood groups are "shaky". Within Treme there are competing interests: those following conventional processes of upgrading historic structures are seen as "gentrifiers who try to push out the local residents". The informant adds, "there's a sort of racial divide that hasn't been bridged very well at all".

In fact, Quarterites feel a strong boundary between the Quarter and Treme, and the physical boundary does exist. As a Treme resident explains. "Armstrong Park[16] you know is sort of a physical barrier". She goes on to point out, however, that many people from Treme work in service industry jobs in the Quarter. Similarly, she points out, "my kids went to school in the Quarter at McDonough 15, so we've spent a lot of time there. So, even then we walked to school from Treme to the French Quarter… so, my neighborhood includes Treme and the French Quarter". Imagining that Quarter residents could say the same is difficult.

A black informant explains that these are the reasons why not many black people live in the Quarter. He thinks it is "both a combination of the psychology of growing up and the rules of segregation and leaving those things behind you and wanting to experience a new life". Besides this he does not see the Quarter as "user-friendly" for black people, making it a "situation where every day is a challenge because the only things that white people from here see in the Quarter are many bad things like the kids tap dancing on the street, all the problems". It seems that New Orleans' black population, too, shares the general idea that the Quarter is a "bad" place. Another black resident links this to the other issues discussed. "I don't know, I think it's looked at more or less as being a place where a…. where a lot of the rich white people live or a place that is not real, it's not a place that has played an important role for black people here in the city. Most of them think that is a place of bad people".

In synthesis the French Quarter is defended by neighborhood groups and individuals both for its historical physical structure and for its diversity of population. Tolerance is displayed toward most groups and open racism is not frequent. Underlying this tolerance is a tension due to the existence of physical and cultural barriers and controversy over whose value should dominate. This becomes particularly notable when linked to a commonly felt resentment toward the black political establishment and its assumed indifference to the concerns of the mainly white and relatively affluent population who live in the Quarter.

Crime: Stricter Enforcement a Threat to Diversity?

Events such as the O. J. Simpson trial have made many white people aware of the profound schisms that exist between different racial groups with relation to the confidence in the US justice system. Black people have constantly complained of harassment by the police but it is only recently that leading US political figures, such as President Clinton and Vice-President Gore, have denounced publicly the widespread existence of racial profiling in relation to those stopped by police. In this context, crime is an issue that moves profound emotions and is bursting with racial tensions. It is not surprising that some of these emerge in an inner-city area like the French Quarter.

For Quarterites, crime is conceived as an issue that threatens the essential pedestrian scale and diversity of the Quarter. The street is their arena and its safety contributes to the liberty of movement that they so value. In the fight against crime, neighbors could be seen as confronting disembedding mechanisms that alienate them from their "place" and threaten their ontological security (Giddens, 1990; 1991). The struggle for the right to the free use of public space can be interpreted as a liberating phenomenon. It represents a defense of the street for unplanned social interaction with diverse groups of people. This has parallels with feminist organizations calls to "Take Back the Night" (see Taylor and Whittier, 1995, p. 178).[17] Nonetheless, in the examination of this issue, the oppositional nature of the neighborhood preservation movements is less easy to sustain. Calls for zero tolerance have an implicit repressive character and may discriminate against certain groups such as blacks and street people, so furthering racial and social tensions. Furthermore, in the treatment of crime, the antagonisms between city government and the residents are further elaborated and expand the tensions implicit in the enforcement of zero tolerance. Residents frequently claim that the police turn a blind eye to infractions, especially by tourists ("out-of-towners"), related to public sex, exposure of breasts and genitals, drunkenness and public urination.

Criticism of the Police and the City – Seeds of Resentment and Racial Tension

Residents and business people in the Quarter are frequently critical of the effectiveness of the police.[18] Interestingly Ophelia points out that the improvements do not all result from the efforts of the New Orleans Police Department. "We have Sheriff Foti's people in the Quarter" and this force has been also collaborating with the local population. She also reminds us of the variety of police departments there are in the city which as well as the City Police and Sheriff Foti's force, there are the harbor, levee and school police forces, aligned with antipathy toward the local politicians in power. Lack of confidence is part of the dissatisfaction with City Government overall. One principal criticism comes from the perception that the police do not enforce zero tolerance in the Quarter. One resident expresses it this way: "I want the Chief of Police to take a look and say zero tolerance. That should apply everywhere. He can't have the nerve to say that you cannot apply it in the French Quarter. It's like saying that my home has to be less than others". Others assign this lack of enforcement on a policy that sees the Quarter as a "free for all zone". The city is blamed and some feel it represents a conflict between black and white interests: "it's just a very short sighted policy on behalf of the city and I think a lot of it is probably racially motivated". A clearer connection is established to the political and racial aspects of the problem, and an explicit criticism of the Mayor. Contradictions, then, are generated between the self-identified tolerance of Quarterites and the call for zero tolerance, which is customarily repressive in nature.

Perhaps the clearest criticism of the police, and of Mayor Morial, can be seen in relation to the December 1, 1996, "Pizza Kitchen" (a Quarter restaurant near to the French Market) killings and the reaction of neighbors to that event. There, one morning, the assassination of three restaurant workers, well known by Quarter residents, unleashed a flood of distress and sympathy. The senselessness of three untimely deaths outraged neighbors. Seemingly spontaneously they organized a vigil in Jackson Square with a march to the restaurant where the deaths occurred to leave flowers. Flowers and messages filled the pavement for many months after the event. This led to the organization of a march on City Hall to express the frustration of residents at the continuing occurrence of such senseless deaths. The problem was that in New Orleans the black population suffers daily this phenomenon. It was not until three white people were killed that any protest was organized. So it was a white march, protesting for the deaths of whites, killed by black young men.

The Mayor interpreted the protest as racially motivated and refused to understand that it contained an expression of revulsion and concern. Instead of meeting the marchers another countermarch was organized, causing a great deal of resentment to people in the Quarter compounding the animosity and distrust residents feel for the Mayor. It was felt the response of the Mayor was unfair. One respondent was particularly resentful saying, "just because, for example the French Quarter decides to march on City Hall because three whites… have been shot it doesn't mean that it is racially based". Another more conciliatory person put the situation into perspective, saying that Mayor Morial "made that a racial thing, it was not a racial thing but it was when he got through with it. He could have come out smelling like a rose on that deal but instead he dropped a great big pile on him. That was very sad, sad for him and sad for us, sad for the city".

The calls for stricter police enforcement, especially the policy of zero tolerance, while aiding the liberty of movement of many, may result in resentment for those who do not appear to fit into the dominant view of who belongs in the French Quarter. It affects also those who do not conform to the socially acceptable image (habitus). A fertile ground for racial tensions is created especially when built upon longstanding perceptions that racial profiling persists.[19] At the same time, the city is criticized, by some, as not wishing to step up police enforcement because it could affect tourist development. This is a further source of racial tensions.

Noise or Music?

Noise, both from music clubs and street musicians, is considered an issue that threatens the residential quality of the Quarter. More than crime, it is a factor that becomes intolerable, from which certain people can find no escape. Frequently, neighbors speak of persons who have been driven out by noise. It threatens the stability of the permanent residential population and is a further disembedding mechanism that alienates them from their "place".

Again, as seen in relation to crime, the struggle against invasive noise can be interpreted as a tendency to exclusion when it affects street musicians of class and racial origins different from the majority of the residents. For some, music is an essential part of the ambience of the Quarter. Seeking its control or elimination represents for them a neglect of the Quarter's cultural history, especially of the black jazz tradition. Some, less sympathetic to street musicians, particularly those who are directly affected by it, want them excluded from the Quarter. Again, the city administration is blamed, this time for not enforcing noise controls because they are felt to be deleterious for tourist development.

Racial Tensions in Relation to Street Musicians

From the perspective of a black woman, active in Quarter affairs, both race and class issues are involved in the discussion about noise. For her, "it ends up being a class confrontation as well (as a race issue), you know. People in general feel that the French Quarter is sort of upper class, and then you have the people that work who are considered to be lower class and unfortunately, in spite of the way we feel about musicians and the historical treasure, they're also sort of part of the lower class. And, artists sort of fall into that category too. So you had this kind of confrontation, you know between people".

Without wishing to argue for any essentialist characteristics associated with race[20] it is interesting to refer to an interchange in a February 1997 meeting to discuss changes in the zoning ordinance that show a different appreciation of noise as a problem:

White woman: "they are all turning into bars".
Black man: "but this is New Orleans not Slidell. Perhaps this isn't the place for you… you want to put the whole city to adjust to you".
Black woman: "if you worry so much about the noise why don't you put up double glazing. I don't mind the noise. It doesn't worry me. I can sleep through it. Lots of musicians need bars for work".
White representative of bars: "there are 300 black bars and only a handful have music zoning. They don't have anywhere to play. There are a lot of musicians here and we want to keep them. They feel very disenfranchised. The music culture of this city is important, you're overlooking a very important segment of the population".
Black man: "and entertainment pays more taxes".

A black woman from Treme gave me her perception of music that displays a more diverse situation in the black community:

Now, in Treme, music is really an integral part of the community. I mean I'm awakened many, many days by my neighbors. I mean, I live around the corner from this musical family, and the little kids, they started out before they had instruments, playing anything that they could find, you know… that's an important part of the musical heritage of the city to preserve. Now, I have teenagers, and they're not as much into history and culture, so the music that these little kids, that I liked to hear in the mornings, my daughters… they don't want to hear the brass band in the morning. I don't mind, you know… there's always a second line, and you open your windows… I enjoy the passings, of the band, and the second line. So, it's an integral part of the community now. We have a lot of people in Treme who have the same adverse feeling about music all the time, anytime, people want to play. Also, people that have chosen to live in Treme, but that's just not a part of the historical heritage that they want to accept and keep perpetuating. But, it's the birthplace of a lot of musical traditions, so, as somebody that's just interested in the history and culture of the city, that's really important for me, you know. I think it's important for the city. I think it's important as a part of our artistic contributions to the world, you know.

This long extract points to the fact that noise can be a problem for the black community but cultural elements need to be assessed. Again racism is a question not just of discriminating against black people as individuals but of not understanding that different groups may have differing interpretations of what is acceptable and what is not. It can revolve around conflicts about whose value system should dominate in decision making. For this woman at stake is a pride in a whole history of her people's culture and tampering with its freedom of expression could be seen as an affront. Music culture is essential to the projection of New Orleans and must be an eternal source of self-esteem for those involved in its evolution.

Neighbors feel that the city turns a blind eye because street musicians are seen also as a tourist attraction and their protection is also part of the economic and development discourse. When asked why he thought the musicians were largely left uncontrolled one person very involved with the noise issue replied that, "they (the

City) see that activity (street performances) as being part of the street life of the Quarter and an attraction to the tourists". Additionally, the elected political establishment is sympathetic to the performers[21] this being an alternative source of employment for local minority musicians and performers. Another noise activist puts the City clearly on the side of the performers. He claims a "hidden agenda" with class and racial elements involved. A racial factor is reintroduced and demonstrates the distrust the Quarter's residents feel about the motives of black politicians.

So again, as with crime, deep animosities are aroused over whose values should prevail and these are presented frequently in racial terms. In the struggle for a peaceful environment for permanent homes, neighbors promote a confrontation that makes them appear insensitive to the cultural, and economic, values of musicians of a race and class different from theirs.

A Black Political Establishment Favoring Development?

Constantly repeated is the perception that tourist development is the prime objective in the Quarter. Residents hold responsible a political establishment, siding with "greedy" commercial interests. Together, these interests give preponderance to a development vision thought damaging to the long-term preservation of the Quarter as a living neighborhood (and as a tourist attraction). The "golden goose" will be killed and the Quarter will become a "Disneyworld". This vision fomenting tourist development can be related to the characteristics of growth machines or development regimes especially regarding the expansion of the tax base and the concern for generating jobs. In the residents' opinion, the political establishment's agenda, at this time personified in Mayor Morial, and a predominantly black political representation in city government, function to placate the black population with short-term job generation. In the city's pursuance of this objective, denizens feel marginalized by the city administration. Despite this Quarterites are sensitive to the political dilemmas implicit in this social reality. Most are antipathetic to racial confrontation but the preservation ideal of mainly white affluent residents, and the black population's need of decent jobs, creates a fertile context for racial tensions.

A White Elite Marginalized? "Everything is Sort of Done Behind Our Backs"

So, though considered by some as a white elite, residents feel marginalized by city government. Some see this as a symptom of black/white confrontation. The common sentiment is that the City does not inform neighbors. It does not care for them. Decisions are taken as though purposefully at the margin of their concerns. Residents complain that only by their constant monitoring can they know what is going on. "Everything is sort of done behind our backs, behind the public's back". This translates into feeling of being "terrified" and "frightened", and of being "closed in" or "run out". The City ignores "hotel expansion, businesses, bars coming in", and "instead of realizing it is against the law, looks upon it as an irreversible tendency and want to simply re-zone everything commercial". Victor speaks of a "free-for-all" which will "kill the heart of the Quarter" seeing it only as "something to exploit". City officials foment this vision. "Many of them tend to think 'if you don't like it, get out'". A feeling of being excluded from decision making and of basic distrust between politicians and residents translates in a neighborhood animosity toward City politicians and their administration.

Residents resent the Quarter being treated like any other area in the city. They label it the "jewel in the crown" of the city, the principal attraction to the millions of tourists that come to New Orleans. For them it deserves special treatment. One vents this feeling. "Well, politicians and government politics say a lot and do very little for this area. We have more than ten million people that come in our village, in our neighborhood, every single year. Then Marc Morial says; 'oh well, the French Quarter is no more important than any other place in the City'. To a degree that's true but not to a very large degree because people don't come to New Orleans to go to Uptown or to Gertown or to any of those places". Another resident is more understanding of the difficulties of asking for special attention in a majority black city. She realizes that other parts of the city have problems, which are felt by their residents to be equally, if not more important. What makes the Quarter distinct is its economic and cultural value. "If you don't preserve this quality, you are going to kill the goose. Now that is as simple as that."

Neighbors, however, feel this as a deliberate policy of neglect. They say that the Mayor and the city administration feel animosity toward Quarter residents. Andrew considers that city government perceives the residents as an elite, demanding privileges not granted to other neighborhoods. What is more, it is seen as a white elite. Politicians manipulate the electorate on the principle of "divide and rule", characterizing Quarterites as rich white. Meanwhile, it is a white elite that sees itself as unsupported by other white elites in the city. As such, it is a minority with no political clout. It feels it neither has a wider support through solidarity with similar class groups nor does it have voting power to influence political action. Furthermore, neither does it have the economic power to compete with commercial and business interests for political support.

Cultural Differences: "They Don't Understand"

Residents complain that the City administration just does not understand the problems of the Quarter. They feel they come from a different culture and have not taken the effort to understand the particular circumstances attached to living in, and defending the neighborhood as an essential part of the preservation of the Quarter. Residents say that politicians "just don't get it, they don't understand about the French Quarter". They do not care if the Quarter becomes a caricature of itself. They do not realize that tourists come to New Orleans because it is "real". The City administration lets it happen through "ignorance". "They are not enlightened as to what this really is and how we should measure this treasure that we have."

This lack of understanding is related to race, and the perception that historic preservation is a white elite concern. The political aspect of the black leaders not being concerned with preservation is linked to what is seen as a general indifference of the black population for the preservation of the built environment. A preservation activist explains that this can be linked to the historical association with slavery. "I think the black governmental leadership. .. they don't understand... sometimes they look at it in terms of the slavery issue and they think a lot of this was built by slave labor, which is not really true. I think a lot of this was built by free men of color who were artisans, craftsmen. But even if it was built by slaves that should be immaterial". Then he refers to what "they" should do. "They should take a pride in knowing that their ancestors did build this and did have the talent and the skills to do it." And that "we" need to educate "those people on that level". Reflecting he admits that this in not necessarily a racial problem either because many white people, too, do not value the historic significance of the Quarter. Again we return to the theme of "us" educating "them".

Beryl understands this dilemma. "If you are a black person, and you are poor and you need a job, to feed your children, saving buildings is not a priority, and I can understand this." Her comments are tinged with an implicit criticism of the black community. She continues; "one of the problems is that we are in a majority black city now and the black community does not see the French Quarter as any part of theirs, although there are blacks who live here, it is not viewed with any great pride except as a place to make money for the black community".

Some people are more explicit in their disdain for the capacity of both leaders and constituents in the black community. One says "there is a great mass of people here that think they cannot make any changes or they rely upon their ministers to tell them, or their politicians to tell them what to do, they are not going to take any independent civil action". She goes on, "the constituency don't have opinions. If the constituency can't read the newspaper, if the constituents are like blocked from having an education by the deplorable level of New Orleans Parish schools then they are not going to have opinions except ones fed to them by their so-called leaders who are only elected because their skin color is the right shade. I mean that is exactly what one has when one has a thoughtless constituency, and not bad people, but people who are not capable of having any kind of defining intellect".

The previous comments show that many residents recognize cultural differences between the black and white communities. Some seem to see this lack of valuation of the historical importance of the Quarter, however, as indicating a general lack of education. Implicit is the assumption that not appreciating historic preservation shows ignorance and cultural deprivation.

There is though awareness that the city does have many serious problems and politicians must attend to them. One political activist expresses this dilemma clearly. He says it is "difficult in a city where you have got people killing one another. I'd leave a meeting at housing development where I'd see a thirteen-year-old child sprawled across the cement, and then I would go into a meeting with twelve little old white ladies talking about the different color of the paint on a building. It's mind boggling". Consequently, as one resident puts it, "preservation has always had to fight being seen as white and rich". He feels that the neighborhood groups are not racist but he thinks there may be that perception and that "may be one of the things that hurts us politically". This is not an easy issue for Quarter organizations because as another person points out, in a majority black city "if you go in and say you're lilly white so where's your black membership, but there are no blacks living here. That's not good, especially for getting anything done, I can guarantee you". So there is an understanding of political priorities.

Neighborhood organizations try to be inclusive of racial diversity but there are elements of tokenism that can make it uncomfortable for black minority membership especially when there is a clear animosity toward the black political establishment, by many of the white majority. This is particularly noticeable in relation to perceptions of Mayor Morial. The words of David put such feelings clearly into relief. "A lot of people here say that the politicians have got the blacks (their political support) and that what Morial really wants is to get rid of the whites." Mayor Morial carries the brunt of the animosity. At one neighborhood group meeting, they started to joke about the Mayor and spoke of painting an effigy of Marc Morial on the pavement and having a "pissing contest" to see who could "piss" on it from the balcony of the apartments. The clear animosity toward Morial expressed by many people was surprising. Why is this so? Is it just because he is from a different class or a different race?

Queenie goes some way toward illustrating possible answers to these questions and explains it by the entrenched attitudes instilled in people in their upbringing that does not fully allow them to accept a black leader. She implies that she feels this is overt racism. To overcome this "it takes a conscious effort to change yourself and see things that you don't like about yourself and figure out a way to change them. A lot of people haven't had to face that". This leads many people brought up in the South that have a "buffer zone of lots of money think that they can just say things about... Marc's a big racist, and it's a black thing". What they do not see, she says, "he is black, maybe not very dark, but he is black and he can't have the black people feeling like he's all for the white people either".

Nonetheless, not everybody believes in this black/white conflict and see it more as a political ploy to set group against group. In fact both black and white politicians have protected commercial interests in the Quarter and both have, at times, been indifferent to preservation interests. As Wendy caustically puts it, "all they did was change the color. We have put different people at the trough that is all".

The black population, too, are including the preservation of historic buildings in their struggle for recognition of their cultural history. In a neighborhood meeting, Beverly Kilbourne of the New Orleans Preservation Resource Center made a presentation in which she spoke of the increasing recognition of the black past in historic built structures. She said there was a need to get African/Americans to understand preservation, making it an important aspect of community revitalization, an aid to fighting crime, generating employment, etc. She referred, also, to a national association of African/American preservationists. In the city an African/American Preservation Council[22] has been formed. It is part of a general trend for the black community to consider architecture with the other cultural elements of black culture, such as music, food, dancing and style. They see it as an important part of strengthening communities.

The Gay Community: Celebrating Diversity

In contrast to the more equivocal attitudes toward the values of the black minority/majority, it is interesting to compare them with attitudes toward another minority: the gay community. Here a uniform tolerance is observable. Overall, respondents show some reservations with the more commercial and extreme gay behaviors – but no prejudice is expressed openly. We observed that respondents were not self-conscious about discussing this subject, their corporeal expression did not show any distaste, and often they brought the subject up spontaneously in conversation.

They generally recognize the gay presence in the Quarter as important. It has become a center for gay culture along with concentrations of gay population and social life in other major cities. Lester traces gay identification with the Quarter from the 1920s seeing its growth as similar to other early gay concentrations such as the North East area of San Francisco, Greenwich Village and West Hollywood. Based upon this historical development local gay men frequently call it "our home base". It has been the "spiritual home" for the development of the gay community,[23] and is most definitely the center of its (commercial) social scene.[24] Gay people feel little prejudice against them there. Although the Quarter may only be one of many options for gay people to live, gay cultural institutions concentrate there (Knopp, 1989, p. 1).

The non-gay respondents talk openly about gay people and are universally tolerant. Edmund White, already in 1983, spoke of his local informant's views that in the Quarter "homosexuality is quite free and accepted even by straights. Gay tourism is undeniably big business" (p. 237). Today, people like Beryl fully accept this sector of the population and in her usual expressive way states that, "I am probably the least anti-gay person that anybody could find, without being gay themselves… the only person's sexuality that's of concern, is my husband's". In fact, it is felt that people who come to live in the Quarter must be prepared to tolerate gays. Victor speaks of people knowing that this is a "major gay haven", and it would be unrealistic to come with the attitude that "well I don't want to have anything to do with the gay community. Because they know that they're moving into one of the most concentrated areas of it".

The bohemian live and let live attitude that we had seem generally in relation to diversity, extends to the gay community and attitudes vary from those that simply see it as none of their business, considering the behavior of others as a private concern, to those that recognize the role that gay people have played in the preservation of the Quarter and stress the value of the gay community to the area. For many it is just not an issue. For others the contact with the gay community has been enriching for their own lives.

Thus, some residents appreciate how the presence of the gay community somehow has a positive effect on their own lives by introducing them to an unknown diversity or giving them access to different points of view, different experiences. Roberta, who came to the Quarter many years ago, expresses something of her acknowledgement of gay people despite a previous ignorance: "I did not know when I came to New Orleans that there were such people as gay people. It was not long before I began to think that there might be, and then concluded that there were. That was a different time, the early fifties. They were nice people and they had the right to their lives as well".

This section, shows that the recognizable efforts of gay people to preserve the Quarter, the establishment of gay social activities together with the investment of gay people in the establishment of business (not necessarily gay-orientated), and the existence of a population which is generally tolerant or learns to be tolerant, all contribute to the formation of a place where gay-lifestyles can be more developed relatively free from the acid gaze of intolerance and from the real dangers of the violence of homophobia. Lawrence Knopp (1989) has indicated that "the French Quarter's long tradition as a relatively open center of gay culture afforded New Orleans gays a much greater opportunity to create integrated gay identities and lifestyles than was available in most other U.S. cities" (p. 65, footnote 2). Various authors have pointed to the importance of certain urban areas as important for the growth of gay identity.[25] As Ken says they are places where "you can recognize that you're not the only one". The importance of a sympathetic environment is very important to gay self-identity. Stanley Siegel and Ed Lowe Jr. (1994), referring to the negative role models often perceived by males who recognize themselves as gay, say that: "Contact with the gay subculture usually provides him with the kind of information he needs about homosexuality to challenge the image he receives and has received from the straight world, and he may begin to see more acceptable possibilities for his future self in homosexual personalities that he finds acceptable" (p. 100). In such circumstances places like the French Quarter can be especially important for development of gay identities, as well as being enriching for non-gay identities.

Nonetheless, there are people who feel there are problems due to the commercialization of gay lifestyles and the behavior of some individuals is offensive. Ken speaks, for example of the fact that "we have reached our peak as far as the commercialism of the gay community, both in terms of the numbers of people and in terms of what we can accommodate". He feels that there are certain problems in areas where there are concentrations of gay bars, exacerbated in the time of festivities, which attract gay people from many parts of the country. A neighbor in this area speaks of the "big gay weekends like Gay Pride, Southern Decadence" when the "people are out in Bourbon, into the street, and will not even let cars pass sometimes. People are fornicating on the street. It is so outrageous".

Final Comments

The preceding arguments indicate that the activism of a minority promoting historic preservation of a diverse environment can be interpreted in different ways. Certainly the idea that this is marginalized minority confronting a dominant majority is clearly inaccurate. The very definition of what constitutes a minority becomes a complex issue. What is even more difficult is to decide if such minorities, because of their status as minorities, have some intrinsic right to be protected and supported in their struggles. Here no clear conclusions can be made because this minority does defend diversity and as such counterposes standardized tourist development. In some ways it appears as a model for the heterogeneity of peoples promised by late modernity. At the same time, minorities defend value systems that negatively interpret some alternative world views and forms of behavior.

Given the delicate nature of this history of racism in the U.S., the situation becomes particularly complex when the minority that defends preservation is one that has traditionally wielded political, ideological and economic power. In this context, preservation of historic structures can be associated with preservation of previously hegemonic white cultural values. While very little open racism is expressed, an underlying current of paternalism suggests that few white people question the underlying assumptions of their culture, that is to say it is the white cultural norms are adopted in relation to evaluations about what is acceptable. The defense of such norms represents an attempt to maintain those forms of capital, particularly social and symbolic capital, associated with historic preservation activism. Additionally, clear physical barriers appear between the Quarter and adjacent black neighborhoods, which make the Quarter unattractive as a place for black people to live, creating real, or perceived, prejudice against them. Finally, the issue is tied to politics and the fact that local politics is dominated by black politicians who could represent groups different to those of the predominantly white residents of the Quarter.

As we have seen before residents were concerned to display tolerant attitudes toward black population but tensions have surfaced in relation to events such as the Pizza Kitchen march. Racial tensions are clearly present in the crime, and noise issues. The development discourse too, is linked to race. In a city with serious economic and social problems, which affect disproportionately the black population, the development discourse centered upon generating jobs predominates.

When these seemingly different visions are placed in dichotomous and oppositional terms, racial antagonisms are maintained on all sides. Achieving compromise in such a situation is difficult, though the majority do not support explicitly such stereotypical reasoning (and in fact are opposed to it). Yet often the reasoning is based upon moral judgements that businesses and politicians are motivated by greed and short-term interests. It is a moral problem where both sides have internalized the negative stereotypes assigned to the opposing sides. Each side abuses the other.

In these circumstances, such antagonisms cannot be eluded but understood, and confronted. The preservation discourse cannot be accepted, a priori as superior. Like all interventions in the city it must be subjected to an ample discursive process that takes into account all interested parties. Inevitably, basic conflicts persist that cannot be resolved. The central conflict presented here of short-term versus long-term interests is a constantly repeated theme typified by the discussion of what should take priority; job creation/economic growth or preservation? The opposition between these two discourses is not necessarily antagonistic but often seems so in the public and private discourses of the actors involved. The question must be asked: is there anything intrinsically valuable about preservation? We think there is, based on the need not to destroy historical structures about whose relevance future generations should have the right to decide. But did the youth of Berlin think that when they tore down the Berlin Wall, or when Lenin was tumbled from the middle of public squares all across Eastern Europe? Buildings and physical artefacts do have profound symbolism for people and are intimately associated with dominant discourses. The French Quarter and its denizens symbolize both exclusion and inclusion. There is great potential, constructed on the tolerance of denizens, for making the Quarter inclusive. In that way it could be justifiably preserved as a proud symbol of diverse cultures, and open to them all. On the other hand, if exclusionary tendencies become dominant, the Quarter will come to symbolize inequality and its preservation could be questioned.

Notes

1   New Orleans is located in Orleans Parish but the urban area spreads over a number of other parishes of contiguous development such as Jefferson, St. Bernard and St. Charles. These adjacent parishes have accommodated a great deal of the suburban development generated, in part, by the outward movement of residents previously located in the Orleans Parish. Consequently, the population of Orleans Parish was in 1990, almost identical to the population in 1940 (almost half a million) although it rose to 628,000 in 1960.

2   Many figures are presented, and exact numbers of visitors is not recorded, but estimates put the figures between 12 and 15 million tourists per year.

3   The principal sources of information are conversations (Kvale, 1966) with thirty-three persons (residents, business people and officials). Data was also obtained from participant observation of events concerning the Quarter during a period spanning 1997 and 1998. The quotations here are taken from this data. Pseudonyms are used.

4   Although he does point to the role of other city-wide organizations.

5   As delineated by Bruce London (1980; p. 87) as a potentially fruitful form of analyzing the process.

6   New Orleans was destroyed by fire on 21st March 1788, leaving very few buildings standing so that "the French Quarter of today, is that Spanish city which rose from the ashes of the French New Orleans" (Saxon, 1988, p. 149).

7   A 1992 report claimed that 78 percent of units in the Quarter were residential (CUPA, 1992, 3–2).

8   At the above 60th anniversary of the Vieux Carré Commission, the panel of preservation activists was generally of the opinion that although preservation is generally given lip-service it is still difficult to achieve especially when it enters into conflict with economic considerations.

9   It should be pointed out that VCPORA is not the only organization in the French Quarter. Reacting to the conflicts, other organizations have been established, that represent particular physical areas or interests in the Vieux Carré. As an introduction, this description, given by an activist, explains something of the main groups in the Quarter. "The biggest organization, for instance, is the Vieux Carré Residents and Property Owners and Associates, it includes residents and businesses. The Friends of Jackson Square is small... most of the members come from around Jackson Square... we have another group, fairly recently formed, which is the French Quarter Citizens for Preservation of Residential Quality. There is the St. Peter's Street group and then you have various business organizations". While creating a greater dynamism considerable conflicts have developed regarding the priorities, which should be pursued by the neighborhood movements.

10 Comments presented on 11th June 1998 at the 60th anniversary meeting of the VCPORA by John Magill, curator of the New Orleans' Historic Collection.

11 Treme is what is considered the first black urbanization in the United States. It is adjacent to the French Quarter, only separated by a four lane local distribution road, North Rampart Street. It is still a mainly black residential area but despite its privileged location has not been renovated like the French Quarter although pockets of "gentrification" do exist, mainly generated by whites.

12 In a meeting to celebrate the 60th anniversary of the VCPORA a member of the audience made some comments using a racist vocabulary. The rest of the people at the meeting became extremely uncomfortable and did their best to stop her from continuing. There were at this time no black people present so the embarrassment was limited.

[13] Ken says that "even as late as that 73 or something like that. Even though laws were passed in the 60s they were not complied with. It took some time for these things to work themselves through".

[14] The boundary between Pennsylvania and Maryland established in the late eighteenth century and regarded as separating North from South.

[15] An informant told me that I had a far too "rosy" picture of this situation and that there is considerable discrimination against black people who dress in a certain more casual way and tend to be younger. Many black people have told me of the constant discrimination to which they are subject in public places, mainly undetectable to white observers.

[16] Armstrong Park is part of the Treme neighborhood and in fact part of it was demolished to make way for it. This created a resentment that has not been forgotten to this day. Nonetheless, it is a potential facility for both neighborhoods. It is, though, often considered unsafe, is fenced, so allowing limited access, and is closed at night, so it is a strong barrier to pedestrian circulation.

[17] Such defense of the street as a safe space at all times is of course important for all minority groups who may be subject to harassment because of their difference.

[18] Some criticism has been tempered by the recent reorganization of the New Orleans Police Department, under Police Chief Pennington. Previously, the Police Department had a deplorable reputation for corruption. A resident says how she now trusts the police department, whereas she says, "I didn't trust the previous one because even if the police were there, I didn't trust them". Thus, generally, there is an acceptance that things have improved.

[19] It should be pointed out that such profiling not only affects blacks but also young people in general, particularly street people, the so-called grungies.

[20] Janet L. Smith (1997, p. 78) points to the dangers of fixing racial identity when employing racial classifications to examine social issues.

[21] Norman also says there are links between the musicians groups and Mayor. "They used to be represented by the Mayor and his brother. There are personal relationships there."

[22] The African American Heritage Committee was formed in May 1997 as an extension of the activities of the Preservation Resource Center.

[23] Lestor also points out the need to distinguish between gay men and lesbians because it is much less a center for women. He indicates that are no lesbian bars in the French Quarter rather the two which exist are located in the Marigny. He again feels that younger lesbians tend to live dispersed for the city. Frances concurs with this opinion saying that, "there doesn't seem to me to be very many lesbian couples, in the French Quarter. I think they tend to have less money than gay men and so it is a less attractive area for them to live in".

[24] This corresponds to the other reports on the location of lesbian women where they were found to favor more dispersed locations, and less expensive property because of generally lower incomes than gay men. See for instance, Valentine (1995) who found that the lesbian community tends to locate in a cluster of homes interspersed with heterosexual homes – without alternative institutions that cater specifically for the lesbians (p. 99). However, other authors while agreeing that lesbians do have lower incomes find that there is evidence of some concentration in "counter culture areas" (Adler and Brenner, 1992, p. 24 and p. 29).

[25] For example Frank Browning (1996) points to gayness as a particularly urban phenomenon where there is a possibility of creating a "gay community", "gay ghetto", or "gay space", the space gay people have carved out for their survival (p. 2).

References

Adler, S. and Brenner, J. (1992), "Gender and space: lesbian and gay men in the city", *International Journal of Urban and Regional Research*, 16, pp. 24–34.

Barthel, D. (1996), *Historic Preservation: Collective Memory and Historical Identity*, New Brunswick, NJ: Rutgers University Press.

Baumbach, R. and Borah, W. (1981), *The Second Battle of New Orleans*, Alabama: University of Alabama.

Bourdieu, P. (1993), *Sociology in Question*, London: Sage.

Browning, F. (1996), *A Queer Geography*, New York: Crown.

Caufield, J. (1994), *City form & everyday life: Toronto gentrification critical social practice*, Toronto: University of Toronto Press.

College of Urban and Public Affairs (CUPA), University of New Orleans (1992), *Changing Land Uses in the Vieux Carré: Managing Growth in a National Landmark District*, New Orleans: University of New Orleans.

Foley, J. (1999), *Neighborhood movements, identity and change in New Orleans' French Quarter*, Ph.D. Dissertation: College of Urban and Public Affairs, University of New Orleans.

Gale, D. E. (1980), "Neighborhood resettlement: Washington D.C.", in Laska, S. B. and Spain, D. (eds), *Back to the City: Issues in Neighborhood Renovation*, New York: Pergamon, pp. 95–115.

Gallas, W. (1996), *Neighborhood Preservation and Politics in New Orleans: Vieux Carré Property Owners, Residents and Associates, Incorporated and City Government 1938–83*, Masters Thesis: College of Urban and Public Affairs, University of New Orleans.

Gamson, W. (1988), "Political discourse and collective action", in Klandermans, B., Kriesi, H. and Tarrow, S. (eds), *International Social Movement Research, Vol. 1. From structure to action: comparing social movement research across cultures*, Greenwich, Con: JAI Press, pp. 219–244.

Giddens, A. (1990), *The Consequences of Modernity*, Stanford, CA: Stanford University.

Giddens, A. (1991), *Modernity and self-identity*, Standford, CA: Stanford University.

Healey, P. (1997), *Collaborative Planning: Shaping Places in Fragmented Society*, London: Macmillan.

Klandermans, B. and Tarrow, S. (1988), "Mobilization into social movements: synthesizing Europe and American approaches", Klandermans, B., Kriesi, H. and Tarrow, S. (eds), *International Social Movement Research, Vol. 1. From structure to action: comparing social movement research across cultures*, Greenwich, Con, JAI.

Knopp, L. (1989*)*, *Gentrification and Gay Community Development in a New Orleans Neighborhood*, Ph.D. Dissertation: Department of Geography, University of Iowa.

Kvale, S. (1996), *Interviews: An Introduction to Qualitative Research Interviewing*, Thousand Oaks, CA: Sage.

Laska, S. B. and Spain, D. (1980a), "Introduction", in Laska, S. B. and Spain, D. (eds), *Back to the City: Issues in Neighborhood Renovation*, New York: Pergamon, pp. viii–xxi.

Laska, S. B. and Spain, D. (1980b), "Anticipating renovators' demands: New Orleans 1980", in Laska, S. B. and Spain, D. (eds), *Back to the City: Issues in Neighborhood Renovation*, New York: Pergamon, pp. 116–137.

Lofland, J. (1996), *Social Movement Organizations: Guide to Research on Insurgent Realities*, New York: Aldine de Gruyter.

Logan, J. and Molotch, H. (1987), *Urban Fortunes: The Political Economy of Space*, Berkeley: University of Berkeley Press.

London, B. (1980), "Gentrification as urban re-invasion: some preliminary definitions and theoretical considerations, in Laska, S. B. and Spain, D. (eds*), Back to the City: Issues in Neighborhood Renovation*, New York: Pergamon, pp. 77–92.

Saxon, L. (1995) [1928], *Fabulous New Orleans*, Gretna, LA: Pelican.

Siegel, S. and Lowe, E. Jr. (1994), *Uncharted Lives: The Psychological Journey of Gay Men*, New York: Dutton.

Smith, J. L. (1997), "Understanding race as a political construct", *Critical Planning*, 4, Spring, pp. 77–96.

Spain, D. (1980), "Indicators of urban revitalization: racial and socio-economic changes in central city housing", in Laska, S. B. and Spain, D. (eds), *Back to the City: Issues in Neighborhood Renovation*, New York: Pergamon, pp. 116–137.

Stone, C. (1993), "Urban regimes and the capacity to govern: a political economy approach", *Journal of Urban Affairs*, 15(1), pp. 1–28.

Taylor, V. and Whittier, N. (1995), "Analytical approaches to social movement culture: the culture of women's movement", in Johnston, H. and Klandermans, B. (eds), *Social movements and culture*, Minneapolis: University of Minneapolis Press.

Thomas, M. J. (1994), "Values of the past: conserving heritage", in Thomas, H. (ed), *Values and planning*, Aldershot: Avebury.

Tournier, R. E. (1980), "Historic Preservation as a Force in Urban Change: Charleston", in Laska, S. B. and Spain, D. (eds), *Back to the City: Issues in Neighborhood Renovation*, New York: Pergamon, pp. 173–186.

Valentine, G. (1995), "Out and about: geographies of lesbian landscapes", *International Journal of Urban and Regional Research*, (19), pp. 96–111.

Wilkenson, T. A. (1985), *Sense of Place and Preservation Planning: An Analysis of New Orleans' Vieux Carré from 1718 to 198?*, Masters Thesis: College of Urban and Public Affairs, University of New Orleans.

White, E. (1983), *States of Desire: Travels in Gay America*, New York: Dutton.

Chapter 6

In the Shadow of Saint Benedict: Leadership, Urban Policies and Ethnic Involvement in a City in Transition

FRANCESCO LO PICCOLO

The really terrible thing, old buddy, is that you must accept them. And I mean that very seriously. You must accept them and accept them with love. For these innocent people have no other hope. They are, in effect, still trapped in a history, which they do not understand; and until they understand it, they cannot be released from it. (...) Many of them, indeed, know better, but, as you will discover, people find it very difficult to act on what they know. To act is to be committed, and to be committed is to be in danger.
James Baldwin, *The Fire Next Time*, 1963

"Different" Inhabitants in "Plural" Cities: a New Reality in Palermo

When describing the contemporary city, a wide range of conditions, living standards, social groups, expectations and needs stand out as symptomatic and salient characteristics. The presence of multiple experiences, processes and people involved making up the urban dimension is to be seen as an established fact: differences (in age, ethnicity, gender, class, religion and culture) are concentrated in cities on various scales and levels of intensity. Roland Barthes (1981), for example, has described cities as "the place of our meeting with the other". Similarly, Richard Sennet (1990) suggests that the urban dwellers are "people in the presence of otherness". Fincher and Jacobs (1998) appropriately highlight in their studies that coping with the theme of difference does not mean indulging in dealing with urban "diversity" (as it happens under rather generic forms in some descriptions of the latest urban phenomena), but rather dealing with networks of power relations, repression and control within the contemporary city.

Recent literature illustrates the new scenarios of difference stemming from such phenomena as international migrations, post-colonialism or the rise of new forms of articulation of society (Loomba, 1998; Sandercock, 2000). Thus, the theme of difference is more and more frequently dealt with in disciplinary debates, starting from the acknowledgement of the fragmentation of contemporary society into an archipelago of "minority" and "plural" groups (Soja, 1989; Sibley, 1995; Scandurra, 1999 and 2001). These groups express specific needs and claim specific rights and benefits affecting the dimension of the city and urban space. The problems arising from this might be tackled in various ways changing from time to time from either repressive or discriminatory to tolerant, inclusive or dialogical-communicative. If we adopt the definition of planning recently given by Healey (1997) as "the management of our co-existence in shared spaces", the subject of this contribution is the way of co-existing in cities of differences, facing the issue of equality and pluralism both in policies and practice.

In the framework of this reference scenario, the analysis of the mutations considered in this contribution is restricted to the increasing population of ethnic minorities in the contemporary city due to recent international migrations. In this particular context, the processes of transformation and initiatives, which Palermo has witnessed in the last fifteen years, are interestingly characterized by analogy and difference. The observation of what takes place in a local context characterized, at the same time, by elements of marginality and innovation may be useful to put forward unconventional considerations about tasks, potential, and a perspective on planning and relevant urban policies in that very context.

Although not strictly comparable with the rest of Europe, due to the distinctive character of its immigration, the foreign presence in Palermo has, in the last few years, taken on significant proportions. Mainly concentrated in the historic center, the foreign communities are extremely differentiated both in terms of the large number of ethnic groups and the manner and time-scale of settlement.

Within the European panorama, the Palermo "case" presents both corresponding and distinctive characteristics. Although simplifying a phenomenon which is certainly more complex, it could be said that what distinguishes the Palermo case – as many others in Italy – from a number of other "tales" of immigration in Europe is its recent character. Unlike Germany, France, the Netherlands and Great Britain, Palermo has been subject to immigration flows which have shown consistency only since the late 1970s, and which became "socially visible" in the early 1980s. More importantly, these phenomena point out a real role reversal. In the last twenty years, Sicily has changed from being a region of emigration to one of immigration (Cusumano, 1976; Giacomarra, 1994; Famoso, 1999). I will not linger on the causes of these phenomena, which have been the object of several, at times contradictory, studies (Lo Piccolo, 2000b). I am just putting emphasis on the specific characteristics of the Palermo case, which is in many ways anomalous in the panorama of European immigration. Through direct day-to-day experience, we find ourselves faced with the most desperate and extreme cases of emigration,[1] a bottoming out in the economic and social characteristics of immigrants with which there is little comparison in Europe (Melotti, 1988 and 1993; Di Liegro and Pittau, 1990; Collinson, 1993; Brusa, 1997).

It is quite difficult to calculate and assess the number of immigrants in the city by means of traditional surveys; and yet through approximate estimations, this number may be considered as significant and able to sensibly modify the composition of the population, particularly in the historic center, also due to the tendency of newly arrived to concentrate in a given place. Not a few studies and surveys report that, still nowadays, no simple and certain outline may be given of the foreign presence in Palermo due to the lack of exhaustive data. As it applies to the rest of Italy, also in Palermo official estimates of this phenomenon are partial and only describe some aspects of the complex reality of immigrants having a lawful residence permit.[2] As a matter of fact, the number of illegal immigrants makes it impossible to objectively estimate the phenomenon as a whole and contributes to a high degree of inaccuracy in data recording and evaluation of processes and needs. Furthermore, another element of complexity is the extremely differentiated character of the foreign population in Palermo, the great number of ethnic groups,[3] diversity of manner and time-scale of immigration and settlement and high mobility of many of the immigrants (Gruttadauria, 1994).

Some research carried out in 1994 took as a starting point the data issued by the immigration bureau of Questura which had recorded 19,758 immigrants, of which 13,211 were men and 6,547 women. Considering that not all of the immigrants were living in Palermo, the total was estimated at 12,000 people, to which some 20,000 more should be added. The places of origin of those 12,000 people living in Palermo were as follows (Gruttadauria, 1994; Tosco, 1994): 43 percent came mainly from six African countries, Senegal, Ghana, Nigeria, Tunisia, Morocco, and Ethiopia; 27 percent were from the Philippines, Pakistan, Bangladesh and Sri Lanka; 12 percent from Mauritius Islands; 18 percent were Iraqi, Algerians, Egyptians, and other nationalities in marginal percentage.

In the light of the complexity and inaccuracy of the various statistical surveys (official and unofficial), as a basis for the considerations and assessment of this contribution, the 1999 data from the aliens' register were chosen. If, on the one hand, the aliens' register does not include unregistered and illegal immigrants (whose number is remarkable, but uncertain), on the other hand it gives an updated and exhaustive outline of the lawful foreign population in the city and its distribution in the various districts, thus offering the possibility to monitor the chronological evolution of the phenomenon and make direct comparison with the data concerning the remaining population.

The foreign population in Palermo, as proved by a comparison of the data of the aliens' register (Attanasio and Giambalvo, 2001), increased by 74 percent from 1992 to 1999. While in 1992, according to the aliens' register there were 9,162 foreign people (5,740 males and 3,422 females) out of a total of 750,121 residents, in 1999 there were 15,931 registered foreign people (9,229 males and 6,702 females) out of a total of 735,975 residents. Therefore, even though on the one hand the proportion of foreign residents, most of whom are non-EC immigrants, is lower than in other Italian cities, as it totaled only 1.2 percent in 1992 and 2.2 percent in 1999, on the other hand the increase in the last ten years is rather significant as well as quite differentiated as per sex, showing 61 percent for men and 95 percent for women.

Yet unlike what happened in other Italian and European cities (Manconi, 1990 and 1992; Paba, 1998; Khakee et al, 1999; Thomas, 2000), the arrival and settlement of immigrants in the historic center of Palermo – as in other areas – has not caused any evident increase in social tensions or explosions of protest. The experience of Palermo is quite different, considering local reactions both at the level of the public opinion and of the City Council administration. The fact is that in Palermo immigrants had settled in a district that was already in a state of decline. We cannot identify in Palermo signs of what has affected other Italian and European cities, that is processes of physical and economic decline, prompted by factors which are extrinsic to local society, tending to be followed by a mobilization of public opinion, which then focuses on a contingent situation in which the immigrant population functions as a catalyst (Somma, 1999, p. 94). At the same time, the City Council's attention and policies for immigrants seem to be quite different if compared to national trends.

In this context the role and tasks of the local scale policies appear, in some ways paradoxically, even more important. Analyzing different cases of leadership and urban regeneration, Judd and Parkinson (1990, p. 14) highlight that, as cities in all the advanced Western nations adjust in different ways to the global economy of the 1980s and 1990s, the significance of local action finally has emerged as a relevant, even compelling topic for analysis (see also Bourne, 1993). The Palermo experiences testify to this, even if in an unusual context and in reference to a particular field of action.

Immigrants in the Historic Center: Ethnic Diversity and Social Polarization

In the last fifteen years the historic center of Palermo has been a place of shelter for immigrants: today it is home to their largest concentration in the city. Although not strictly comparable to the rest of Europe, due to the distinctive character of its immigration, in the last few years the foreign presence in Palermo has taken on significant proportions. Mainly concentrated in the historic center, the foreign communities are extremely differentiated both in terms of the large number of ethnic groups and the manner and time-scale of settlement.

A great proportion of immigrants live in the historic and decayed centre.[4] They have settled in an area already depopulated and in decay (Cannarozzo, 1990 and 2000; Lo Piccolo, 1996 and 2000b). However, in spite of decay and depopulation, Palermo's historic center (250 ha in area and about 27,500 inhabitants out of the city's total of 750,000), retains the strong identity of a great capital, underlined by the richness and variety of its architecture. Moreover, it continues to have a recognizable role in the wider urban context. So this area is characterized by two main phenomena – on one hand a progressive depopulation and, on the other, a classic history of immigration (Cannarozzo, 1996 and 2000; Lo Piccolo, 1996).

In recent years immigrants have begun to move into buildings long abandoned and generally unfit for habitation, where they have a precarious existence (Lo Piccolo, 1996 and 2000b). The exploitation of the vulnerabilities of this new marginal population has inevitably led to further privation.

Most often, owners have left these buildings behind because they were unfit for habitation or in shaky condition. Regardless of this, monthly rents range from as much as 300,000 liras for a single-room lodging to one million liras for four-room flats. Dwellings are usually made up of small rooms having almost no lighting nor ventilation and insufficient sanitary facilities. On-the-spot observation shows that immigrants prefer to lead their life outdoors rather than indoors, their lifestyle being still linked to their country of origin and dwellings overcrowded and unfit for habitation. People coming from Northern Africa and Nigeria have the poorest standards of living, usually with groups sharing single flats because of poor wages. Anyway, these immigrants, regardless of their culture and country of origin, very often live in dwellings having very low standards of habitability.

So far, the historic center has not been affected by any process of gentrification, the consequence of the condition of its properties and general decay as well as the lack of public investment. Recently some indicators seem to show slow changes in that respect, and it is not at all to be excluded that gentrification processes will start in the near future (Cannarozzo, 1999).

When considering the demographic structure of the residing population, the historic center has today various elements of singularity with respect to all the other areas of the city as a consequence of diverse and in a way contradictory processes, some of which have started quite recently while some others have been going over a much longer period of time. Such phenomena may be briefly explained by a process of depopulation and social polarization towards lower strata which started in the early post-war years and continued with alternate rhythms and phases through the early 1980s, by a process of "settlement" of migrated populations starting from the very first waves of migration (strengthening of social polarization toward lower strata and of sprawling social marginality) and by a process of "return to the center" of affluent classes (starting from the late 1980s) with quite specific and characteristic family compositions, similar to forms of gentrification which took place in other Italian and foreign urban realities. In this regard, an analysis of the demographic structure may turn out to be quite useful, provided that co-present and "contradictory" phenomena are taken into consideration.

First of all, the prevailing number of foreign residents, even if totaling only 2.2 percent of the overall population, in the historic center reaches 15.7 percent. The number of elderly people is larger in the historic center than in other areas of the city, while age group 40–64 is lower. At the same time the gap between the percentages of residents younger than 15 years (17.9 percent of total) and age group 15–24 (12.3 percent) is more than five percent, which does not apply to other districts of the city (Attanasio and Giambalvo, 2001), where the difference is much smaller. That may be explained by the presence of a second "generation" of rather young foreign residents, whose majority is still under 15 years of age, and the scant number of new-born among Italian residents.

The historic center is an exception also as regards the composition of families. Some research (Attanasio and Giambalvo, 2001) show that the historic center is the farthest from average values. Most interesting are the data concerning single-member families, which can be explained by processes of aging of the population due to the population flight in the 1950s and 1960s, and also by the return of new affluent inhabitants: Italian single males younger than 40 make up 23 percent in the historic center compared with nine percent in the rest of the city.

By comparing these data with the citizenship status, particular emphasis can be put on the strong concentration of foreign, male, not elderly citizens in the historic center, which is also clearly inferable from an intuitive point of view. As a matter of fact, if we consider that in Palermo 73 percent of foreign single-member families are made up of males (compared with 31 percent of Italian families), the cross-analysis of such variables as type of family, citizens and district shows that a large number of foreign single-member families live in the historic center (40 percent compared with seven percent elsewhere in the city) and are mainly made up of males.

These data show a standstill in processes of population flight and depopulation, followed by a modest but ever increasing return of inhabitants (Cannarozzo, 1999).

The decay of the area has, in the past, brought about a decrease in the value of property and rents, despite its central location. Historically, immigrant communities have chosen to settle in areas such as this for precisely these reasons of low-cost housing and location. In recent years, initiatives for the revitalization of the historic center have been taken. Doubts are raised about the survival of the ethnic communities, which risk being swept aside by re-development which will substantially change the appearance, structure and function of the whole area (Lo Piccolo, 2000b). The envisaged future redevelopment work will raise property prices with a consequent rapid loss of low-cost housing. In the near future, the historic center will be the contested terrain of the next round of urban redevelopment in Palermo, the remaining frontier. Even if, by now, it is quite unlikely to consider any redevelopment activity as "ethnic cleansing", such an increase in prices will inevitably trigger off a process of expulsion of the lower income groups.

Appropriation of Space: the Emergence of New Geographies

Migratory processes, and consequent settlement and identification with any given place, are inevitably long drawn out, often contradictory and in any case non-linear. They involve not only the entire life-span of the individual who migrates but also subsequent generations (Castles and Kosack, 1973; Miller, 1981; Castles 1984 and 1989) and constitute a true collective action which produces important social transformations and consequent changes (which are also physical) in the new place of settlement (Castles and Miller, 1993; Collinson, 1993; Friedmann, 1995).

As Massey (1991, p. 227) reminds us, places are constructions emerging from the intersection and interaction of social relations (local) and wider social processes. These constructions may entail forms of conflict and mediation affecting not only the nature of physical places (and the meaning ascribed to them), but also the physical and symbolic outline of their borders. The role and outline of places and borders involve a wide range of meanings arising from the complexity of social relations and co-present social processes (McDowell and Massey, 1984; Massey, 1991; Thomas et al, 1996). Therefore, migrations have a strong and direct bearing on the outline, transformation and use of spaces, and of urban spaces in particular. In this regard the theme of the identity of the city acquires new, further significance (Lo Piccolo, 1995; Attili, 2001), both in terms of the variation in the identity expressed by the city with its new ("different") inhabitants, and from the point of view (opposing and complementary) of the recognition of the urban spaces by ethnic immigrant groups. This recognition does not have individuals as its exclusive object but also the spaces and forms of the city (Indovina, 1991; Sibley, 1995; Lanzani, 1998), "negotiating", in the city spaces, "different" forms of interaction and co-habitation (Bird, 1992). These phenomena can nowadays be recorded in Palermo as well, particularly in its historic center. Some considerations can be based upon the processing of the data concerning foreign residents and direct monitoring of ongoing phenomena.[5]

The number of people living transitorily and informally is quite modest with respect to this "new ethnic geography" in such cities as Milan, Florence, and Rome. Multiple factors ranging from a high level of tolerance, light control and law enforcement, widespread conditions of decay and depopulation favored a rapid and far-reaching settlement yet with strong forms of social marginality and exclusion. However, the settlement of new inhabitants has not yet given birth to real ethnic districts, as in the case of concentration and ghettoisation. These "new geographies" appear to have a more complex and fragmented articulation, which, similarly to other Italian cities, creates a thick network of "urban micro-colorations" which, in its turn, slowly forms new physical and social spaces (Paba, 1998; Perrone, 2000).

With reference to the whole historic center, most of the housing is concentrated along its main streets (Via Maqueda, Via Roma, Corso Vittorio Emanuele) and in the historic marketplaces (Vucciria, Ballarò, Capo), which have traditionally been places of exchange and meeting points of "alien" cultures. As regards Corso Vittorio Emanuele, a higher concentration of dwellings is found in the area between Quattro Canti and Porta Felice, whereas lower concentration can be found in the southern part of the historic center, probably due to the presence of utilities and restored buildings. The lack of real ethnic districts is found out by analyzing the number and distribution of North African immigrants, the most numerous group both in absolute percentage and with reference to the historic center alone. Immigrants coming from Morocco (268 men and 40 women, 308 people out of a total of 1,135 individuals in the city) are mostly located in Mandamento Tribunali (148 residents); they are evenly distributed over the whole area and co-habit with other ethnic groups. Recorded data show that the number of men is considerably higher than the number of women: consequently there are a small number of families and the area feels overcrowded.

Tunisians (557 men and 212 women, summing up to 769 out of a total of 1960 people in the city) are the most numerous ethnic group over the entire city territory. Unlike Moroccans, Tunisians tend to concentrate and dislike the co-habitation with people of other nationalities, like, for example, the large groups living in Mandamento Castellammare (201 residents). In spite of this, in Mandamento Tribunali (227 residents), Monte di Pietà (185 residents) and Palazzo Reale (156 residents) they are more evenly distributed and mixed with other ethnic groups, as it is the case for Via Porta di Castro and the area of the historic Capo marketplace. Many families live there and consequently houses are less crowded.

Immigrants from Bangladesh (449 men and 151 women, summing up to 600 out of a total of 1,299 people in the city) occupy a few dwellings which are consequently overcrowded; as a matter of fact they are used to living in numerous groups or in enlarged families. They do not share housing with other ethnic groups with the exception of immigrants from Sri Lanka.[6] They are mostly concentrated in Mandamento Castellammare (127 residents), where a place of worship also exist in Via S. Basilio. A large number of them also live and run food and service shops around Via Roma, Via Fiume and Discesa dei Giudici. As already made clear in literature (Tosi, 1998; Lo Piccolo, 2000a), it is not easy to refer to settlement processes of ethnic groups as to homogeneous phenomena as per characteristics and structure: because of the institutional and interpersonal relations developing between old inhabitants and newcomers, the forms of settlement may vary considerably as a consequence of political and cultural contexts which may stretch from out-and-out acceptance of citizenship rights to the most extreme and violent forms of discrimination and socio-economic isolation, sometimes even through segregation. Some elements or processes act as "repulsive", some others as "attractive", taking forms which are undoubtedly more complex than those dictated by the rules of the real estate market and housing offers. Some of the indicators through which one can analyze these new settlement processes are family and community ties, informal networks, welfare and service centers and the development of ethnic-related micro-economies.

In the light of our interest, the Care and Support center of Santa Chiara is absolutely noteworthy as it gathers a large number of non-EC immigrants, mostly coming from Central Africa and enjoying, on a continuous basis, the many services provided by the center. In this regard, most of those coming from Ghana (286 men and 190 women, summing up to 476 out of a total of 997 in the city) live in Mandamento Palazzo Reale (234 residents) and Mandamento Tribunali (134 residents) and tend to gather in large communities. Immigrants from the Ivory Coast Republic (91 men and 65 women, summing up to 156 out of a total of 352 people in the city) very often have no family ties and live near the center of Santa Chiara (82 people live in Mandamento Palazzo Reale) along with immigrants from Senegal. According to official data, the presence of women is almost unrecorded and the data obtained on the basis of "regular" surveys give an underestimated outline. The Nigerian community (28 men and 16 women, summing up to 44 out of a total of 92 people in the city) almost entirely live in Mandamento Palazzo reale (32 residents); these data too are unreal and underestimated. As a matter of fact the number of Nigerian men and particularly Nigerian women is much larger in the historic center: there are about 400 of them, of whom more than half are women.

Therefore, just as the center of Santa Chiara acts as catalyst and reference point, thus affecting to a certain extent the geography of new settlements of some communities coming from Central Africa, other ethnic groups choose their residence on the basis of the concentration of work activities (shops and restaurants). Almost the entire Chinese population (27 men and 17 women, summing up to 44 out of a total of 213 people in the city) live in Via Bandiera and Via Cavour (24 residents in Mandamento Castellammare), where there are also Chinese shops and restaurants. Similarly, the large number of immigrants from the Philippines in the historic center, almost entirely working as house helps (21 men and 27 women, summing up to 48 out of a total of 858 people in the city), is per se reduced and concentrated along the Volturno-Cavour axis, near the nineteenth-century district and uptown where upper and middle classes live.

It is clear that the construction of "new geographies" is not only conditional upon housing requirements alone. More generally, as regards commercial activities run by immigrants in Palermo, beside the traditional sector of restaurants (Tunisian, Chinese, Tamil cuisine and others), in recent times new retail stores have opened specializing in handicraft or foods from immigrants' countries of origin. In particular, in the historic center a certain number of businesses have been set up by the communities from Central and Northern Africa and Bangladesh; also some "telephone centers" have been set up by immigrants (Bangladesh and Ivory Coast Republic) to provide international telecommunications services. Therefore, like in other Italian cities, (Granata, 1998; Lanzani, 1998; Paba, 1998) a micro-network of ethnic-related economies develop, which transform some city spaces and the way these spaces are ultimately used.

Expressed Needs and Unexpressed Demands: Role and Effectiveness of Institutional and Non-Institutional Actors

Within this articulated and complex framework of people, lifestyles, needs and settlement patterns, institutional and non-institutional solutions in terms of support, care and services are quite heterogeneous and discontinuous, but at the same time rather "fertile" and require a detailed analysis in order to give perspective to evaluations and considerations.

As in most Italian cities, religious associations and institutions play a key role in providing support, care and services. Sometimes, this role is limited to traditional forms of support and care as it is the case for the center run by the nuns of Mother Teresa of Calcutta, located in piazza Magione (distribution of food and clothes); in other cases activities and services have a broader scope and do cover better organized areas.

This applies to the center of diocese Caritas "Agape", located near the church of Santa Chiara whose support and care role is focused on the social and health sectors: everyday 30 to 40 people resort to the general day-hospital (internal medicine, children's medicine, specialized medicine). The S&C center for non-EC citizens carries out a similar role as regards first aid, integration and training. The center was founded in 1989 by the diocese Caritas and has always been directly run by the Palermo diocese through the confraternity of Santa Maria del Soccorso and is located in the court of the Gancia monastery.

Within the above-mentioned scenario, the Santa Chiara center has been acting as the leading meeting and service point for over ten years. Moreover, the center has also waged important civil battles, such as the fight against juvenile exploitation and pedophilia in the district of Alberghiera where it is located. The center of Santa Chiara offers a wide range of social, health-care and support services along with the opportunity to attend religious and cultural events, mainly to communities from the Ivory Coast Republic, Ghana and Burkina-Faso.

The S. Chiara parish, the main assistance and social center for immigrants in Palermo, allows – even if in a sort of "non-official" form – the practice of different religions: a small mosque is hosted in a room near the church while Hindu ceremonies and African animistic rites, including funerals, are celebrated in the court. This testifies, besides the undoubtedly significant role of the volunteer and religious sector in the supply of social and cultural services, a good level of integration and religious tolerance, even if mainly stimulated by the individual will and open-minded personality of the parish priest Don B. Meli, who was recently appointed advisor to the City Council for immigration issues. On the other hand, to give a more complete picture of the context, it must be noted that the Catholic Church does not allow the celebration of other religious faiths in its establishments, and that Don Meli was consequently invited by his superiors to stop his "open door" practice of religious tolerance.

Religious needs are, in fact, often not satisfied, as there is a lack of places of worship for most of the ethnic minorities living in Palermo. For example, there is not any Hindu temple, although the city has a community of at least 4,000 individuals. The Tamil and Mauritian communities share the same limitation, consequently being forced to use private houses for their religious practice.

The role of Trade Unions (CGIL and UIL) is limited to helping immigrants through the various steps for getting official papers (lawful residence permit, family reunion, change of residence, registration to employment bureaux) and supporting/ advising immigrant workers (lawsuits, job search), while, on the other hand, S&C centers run by the representatives of foreign residents and associations of immigrant communities are making quick progress towards competitiveness and wide-scope services. Some of these associations are quite sound and active and mainly made up of immigrant women (let's remember Titina Silà Association, Union of Filipino Workers – whose spokesperson, Rizalina Santiago, was chosen among the representatives of the consultative assembly, the Association of Tunisian Mothers, whose main sector is recreational and after-school activities, the Association of Immigrant Women and "Vehivavy" Families gathering women from Sri Lanka, Tunisia, Madagascar and Algeria). Some more centers have temporary features and are characterized by elements of ethnic or religious identity (the Catholic Association of Ghana, located within the church of S. Nicolò in the district of Alberghiera and other associations of the Mauritius, Cape Verde, Tamil, Islamic, Palestinian, Bangladeshi, Ivory Coast and Ghana communities) which mainly carry out recreational, cultural and religious activities.

Some communities show particularly enthusiastic commitment in providing educational services of mother language and culture; among them the Tunisian school (promoted and funded by the Tunisian consulate) located in Discesa dei Musici, the Tamil school (run by the Tamil community and hosted during afternoon hours at the State secondary school "G. Daita", via Fiume, in compliance with an agreement signed with the municipality), the Tunisian day nursery run by the Association of Tunisian Mothers having its seat at the State secondary school "E. Ferri", via Discesa dei Giudici.

The work and activities of some of these immigrants' associations is integrated with the activity of non-profit associations, project managers and other significant initiatives, such as CEPIR (the center for fostering integration of refugees), founded in 2000 and run by CISS (International Co-operation between Southern Italy and Southern World) on behalf of the provincial government of Palermo and which aims at fostering support and integration of refugees in the metropolitan area of Palermo.

This whole body of people involved, initiatives, and services presents an image of an extremely dynamic social framework with a high potential for self-organization and integration with other actors (particularly in the voluntary sector), but also showing high rates of marginality and exclusion. As a consequence, within this particular scenario, it is not simple, but nevertheless necessary, to understand what the actual needs are and what kind of initiatives should be undertaken.

The demands put forward by immigrants are also acknowledged by a great number of studies (Gruttadauria, 1994; Tosco, 1994; Lo Piccolo, 2000b) and surveys required by the municipality to determine what the marginal demand is,[7] and reflect different needs and expectations characterizing the structure of the foreign presence in the city. The main deadlock is that these demands are basically unexpressed by individuals who are reluctant to express their own needs and, in a broader sense, to "unveil their own life-story". As a matter of fact, these very people generally do not trust public administration on the grounds that, in recent years, deeds have always been partial and poor. On the contrary, the latest City Council administration succeeded in winning consensus thanks to its far-reaching initiatives for the recovery of the historic center and its support for immigration policies, thus bringing about renovated hopes in Palermo (Cammarata et al, 1996; Cannarozzo, 1996 and 1999; Lo Piccolo, 2000b).

If we look at this bright and heterogeneous panorama, which is nevertheless not broad enough with respect to the needs expressed by ethnic communities, the policies undertaken by the City Council administration turn out to be only partially adequate, being quite sound in programs and goals but poorly effective in integration/correspondence with needs, initiatives and desires of ethnic communities.

In the Shadow of Saint Benedict: The Progressive Agenda in the City Council Administration

In recent times the Municipality has had a new attitude towards immigration, aimed at creating a unified program that values the different cultures present in the area and removes social, religious and economic barriers preventing foreign residents from enjoying their rights. To achieve this goal, the Municipality set up a general project acting as a framework for the various initiatives, both those in progress and those still under discussion, pointing out the aims pursued by the Municipality, its specific programs and available and necessary funds, as well as timetables (Cammarata et al, 1996).

The guidelines of the City Council program are as follows: whenever possible, avoid the creation of specialized services for immigrants, while improving and favoring access to those structures and services already in operation; work in close co-operation with associations, community groups, agencies and private volunteer groups that are active in this field.

In the *communications sector*, the main goal was to improve the network of relationships and exchanges that existed in the city, through the expansion of the multi-lingual information desk at the Office of Public Relations and the continuation of the decentralized Registry Office at the S. Chiara center of emergency assistance.

In the *health and social services sector*, while keeping in mind the limitations imposed by national and regional regulations, the initiatives included: the expansion of the multipurpose center Al-Khalisa, via Scopari, which operated as a consulting and planning center, by the addition of an archive for regulations and statistics, the collaboration of foreign personnel, a secretarial service and a legal consulting service; the creation of a women's center, should the desire be expressed by immigrant women in the study carried out by the LIA Project (Local Integration/ Partnership Action), an arm of the EU, in co-operation with immigrants group and NGO; the soon-to-be-opened municipal center of emergency and mid-term assistance, located in the buildings of via Chiappara al Carmine.

In the *cultural exchange area* the activities were mainly oriented towards the organization, management and funding of shows, film festivals, games, meetings, ethnic and multicultural festivals. A project was prepared for a soon-to-be-opened multicultural center with a multilingual library including newspapers and magazines.

In the *sector of schools and education*, many projects and initiatives were devoted to promoting and developing the culture of origin in Palermo's schools. To this purpose, classrooms were made available for after-school classes run by immigrant groups; adult literacy courses and Italian language courses for minors with language problems or learning deficiencies were carried out in co-operation with immigrant or private volunteer associations. Furthermore, "A multiciplity of stories in Palermo", a multicultural education project,[8] has been proposed for the second consecutive year by the Board of Education (*Assessorato alla Pubblica Istruzione*).

According to this program, the Municipality rejects the logic, which is also shared by some recent state legislative measures, that considers immigrants as a "problem" to be faced and solved mostly in terms of public order; in Palermo, a city which has always been a melting pot of races and peoples, the presence of so many people from non-European countries must be seen as a "resource", a tremendous opportunity for the meeting of cultures and for a mutual enrichment (Cammarata et al, 1996, p. 3).

As a consequence, the tendency should be avoided to plan services, structures and interventions for immigrants only, which would isolate foreign communities from the city's social context; meeting and exchange opportunities should instead be promoted among the different communities. According to the Municipality program (Cammarata et al, 1996, p. 3), the best service the Municipality can offer to foreign citizens is to start or develop primary services and facilitate access to them for all citizens, regardless of their nationality; to attain this goal, the Municipality shall remove all obstacles, including those deriving from regional laws, preventing the full use of city services, and shall widen information networks by establishing new offices in various city locations according to a functional distribution pattern. This is not meant to rule out, at least at the beginning, the need for a number of ad hoc structures and services, mainly targeted at immigrants; but the intent of the administration is to draw attention to all the needy sectors of the population (including immigrants), so that all services can adequately meet the needs of those groups.

Special attention must be devoted to family issues: most of the foreigners in Palermo now live in households. It is thus necessary to start and improve at first those services, which can promote family integration as well as wider access to city structures. Furthermore, if we just think of the fundamental role women (and children) play in family integration, we can easily understand the importance of "women's issues" (Cammarata et al, 1996).

Even if we agree in general with the good intentions of this program, some points raise concerns, notably emotional assertions of tolerance and solidarity, which hide the conflictual nature of the phenomenon. One basic misunderstanding has its origins in a comforting vision according to which the encounter between different cultures and ethnic groups in the context of one society will necessarily bring about effects, which are beneficial and homogenizing.

For example, with reference to the issue of faiths and cultures, the Municipality is generically aware of the remarkable variety of religions professed among the city population and thus engages to promote the free expression of any religious faith, also contributing to the creation of places of worship. This assertion can be agreed upon on the one hand, but it is too generic on the other, putting aside possible tensions amongst different ethnic groups, especially in relation to allocations and priorities in the distribution of the (limited) funds. In this respect, the Municipality program fails to consider the conflictual dimension either amongst ethnic groups or between these very groups and the Italian population, and draws down a multicultural policy that might be seen as acceptable by everyone.

Even if Palermo has a low percentage of racial violence acts in comparison with other Italian cities (see Caritas, 1994), this aspect cannot be neglected. Unfortunately, the issue is not only one of racism versus tolerance, nor does a multi-ethnic society turn out to be a kind of neo-folkloric entity, eating couscous and dancing exotic folk dances into the night. As it has correctly been observed, ethnic-cultural forces are among the strongest of our age, and the conflicts which such forces give rise to in the absence of adequate institutional frameworks are among the most persistent and least negotiable (Pacini, 1989; Brusa, 1997). Moreover, we must bear in mind that the different demands formulated within a city are expressed by (social and/or ethnic) groups, which, quite often, do not represent general interests, but partial and sometimes even opposing interests and which consequently can clash rather than co-operate.

Reading between the lines of the Municipality program, it is possible to detect a certain degree of awareness of such issues; in fact, the program underlines – in its general comments and guidelines – the necessity to proceed not just in terms of solidarity, which concerns individual behavior, but rather in terms of the affirmation and recognition of rights, also by means of procedural forms of legitimization and participation. As we shall discuss in the following paragraphs, the recent election of immigrants' representatives, as a result of the firm political will of Mayor Orlando, serves this purpose and positively confirms some statements of the program. The issues of political representation and the right to citizenship are naturally involved, and still unresolved with reference to what remains the basic issue – the relationship between minorities and the majority (Indovina, 1991; Pinna, 1993; Holston, 1995).

According to their program, the Mayor and the City Council appear to be truly sensitive to immigrants needs. Even if the relatively high degree of tolerance and acceptance shown by the population of Palermo can be explained by many social and cultural factors (some of them not to be considered entirely in positive terms), the policy undertaken by the local government administration also seems to have contributed to this. On the other hand, there are more words than deeds, as it is the case for the frequent and quite instrumental use of a "rhetoric of tolerance". A minor but interesting example is the "elevation" of a hitherto unknown patron saint of the city, Benedict (1526–1589), who accidentally (but now significantly) was black, to the same prominence as the generally accepted and honored patron saint Rosalia (1125–1160), who (not less significantly) was white. This could be a constructive gesture but only if followed by actions which address the day-to-day problems of immigrants. Furthermore, the rhetorical use of this image (these couple of saints, different in race but "working together", in alliance and harmony, to protect the city) seems to gloss over divisions and conflicts, as well as existing inequalities (Lo Piccolo, 2000b).

Notwithstanding the positive political attitude towards immigration and integration[9] and above-outlined articulated program (Cammarata et al, 1996; Lo Piccolo, 2000b), the social policies undertaken by the municipality, in the light of today's results, have turned out to be limited in time[10] or poorly effective, particularly when involvement of immigrants' communities was negligible and/or when measures were conditional upon EU fixed-term allocation of funds. This is the case for the multipurpose territorial social center Al-Khalisa, located in the historic center, Via Scopari. The center, run by the municipality itself (Board for Social and Healthcare Activities and Immigration issues), was opened some years ago and worked successfully during just the first two years of operation, while later immigrants' interest in the center slowly dwindled. Nowadays the center provides services not aimed at immigrants.

Much better results were obtained through initiatives and projects jointly fostered with immigrants communities, on one hand, or many other social institutions, of which only a limited number are related to immigration issues on the other. Project Al-Giazirah belongs to the former case: it is financially supported by the municipality of Palermo and implemented by the Italian Red Cross. The project enjoys the money provided by the territorial plan for intervention for children and youth (in compliance with law 285, 1997) and aims at protecting minors at risk by offering them the possibility to grow up safely and be educated.[11] It must be underlined that today the project, whose person in charge is Sirus Nikkho, from Iran, who was elected among the representatives of the consultative assembly, has broadened its scope: at present not only minors at risk are successfully taken care of, but also minors belonging to any social origin.[12] S. Anna S&C center belongs to the latter case. The center was founded in 1997 and is in the Kalsa district: it is a multi-service center aimed at families and more generally at those living right in the historic center, including immigrants. This project is promoted by the Palermo municipality and European Union – European Fund for Regional Development (Project Urban) and provides for a series of integrated interventions jointly carried out by municipality workers and volunteers.

The projects which are already operational or which will soon be started include the project for a temporary lodging center at Via Chiappara al Carmine, which was initiated when law 40/98 was first introduced. The project is now awaiting start-up money and aims at the recovery of the residential function of two buildings located at Via Chiappara al Carmine.[13] Also the creation of a permanent multiethnic center is envisaged by the Palermo municipality (supposedly in partnership with provincial authorities) and which will be entrusted to non-profit associations. So far locations have not been chosen: some suggest the railway pavilions of the former Lolli station, other indicate one of the pavilions of the Zisa cultural site. Needless to say that the representatives of immigrants' communities, for quite clear reasons, flatly refused the suggestions of a temporary structure (a big tent) near Via Archirafi. At this very stage, the representatives of immigrants' communities are invited to attend meetings and express their needs and views but have no say in the ultimate decision-making process.

The formal incorporation of the non-profit sector has been made possible through consistent and persistent public leadership. The first attempts carried out by the City Council demonstrate once again that participation is a slow and difficult process consuming time and money which is considerable and often seemingly out of proportion vis-à-vis the results obtained: it is an "imperfect" process, just for the lack of a direct causal link between effort and results, as well as for the lack of a guaranteed relationship between the demands expressed and real demands (Arnstein, 1969; Fisher and Kling, 1993).

Reference to ethnic plurality burdens this issue with still more problems, as a consequence of the heterogeneity of the groups and of the lack of connection between the different individual identities and the identity of the ethnic group, which can in no way be taken for granted (Prashar and Nicholas, 1986; Ratcliffe, 1998; Sandercock, 2000). Each ethnic minority cannot be considered – in cultural, social, anthropological or political terms – as a homogeneous and undifferentiated group, fossilized in time, but as containing different specificities and individual characteristics, as well as several levels of integration/non-integration which do not make the question of "who represents who" in the participative process any easier to solve (Lo Piccolo, 1999).

This clearly appeared in Palermo when consultative meetings were held with ethnic communities for the project of the multiethnic centre, as in some ethnic groups disputes, which arose on the selection of representatives; such a problem also took place in developing projects in the health and social service sector for immigrant women. The unsuccessful management of the municipal multipurpose social centre "Al-Khalisa" seems to be in part due to the lack of correspondence between communities and those few individuals directly involved in running the centre: in many cases ethnic communities have seen the latter as "people working for the City Council and having benefits from this", consequently not recognising them as their representatives.

The recent experience of the Advisory Council of non-EC foreigners, refugees and asylum seekers, which is quite a remarkable hint of the political trend followed by the municipal administration,[14] and which is a clear and positive exception over the whole domestic panorama, is evidence of how these problems still persist.

Your Vote for a Colored Palermo: Consultative Elections of the Representatives of non-EC Foreigners, Refugees and Asylum Seekers

On 10th October 1999 the City Administration of Palermo called the consultative elections of five representatives of non-EC foreigners residing in Palermo and of one representative of the (political and not) refugees and of those who obtained asylum. Foreign citizens residing in Palermo since 15th June 1999, refugees and those who had obtained asylum and whose status relevant authorities had acknowledged were given the right to active and passive vote. They had to be over 18 years of age by 9th October 1999. The citizens with double nationality (Italian/foreign) are not entitled to vote, while the citizens with double foreign nationality are.

The candidates were nominated and registered at the Assessorato alla Persona, la Famiglia e la Comunità (City Council Department for the Individual, the Family and the Community). Only the candidates backed by no fewer than 30 supporting signatures were accepted.[15]

Six foreign representatives (one of whom, having the status of refugee or getting asylum, representing the refugees) were elected in accordance with the following procedure:

- the vote has to be expressed with a single preference on two ballot papers (one for the foreign representative and one for the refugees' representative). The five non-EC citizens' representatives and the representative of the refugees and asylum seekers who have obtained the greatest number of valid votes will be elected;
- no more than one representative per nationality of origin will be elected, with the exception of the representative of the political refugees. If amongst the first five elected representatives more than one share the same nationality, the one who has reported the greatest number of valid votes will be declared elected and the vacant seat (or seats) will be assigned to those who follow on the name-list as per number of votes up to five representatives of different nationality;
- the representative elected by the refugees will be the one who, among the candidates of the relevant list, has reported the greatest number of valid votes. In this case, considering the special status of the refugees and those who obtained asylum, the nationality of origin of the elected representative will not be considered for the allotment of the seat;
- only the ballot papers, which show one preference among the candidates for the foreign representatives, will be considered as valid.

About 1,500 immigrants voted: this may seem quite a small number, in comparison with the number of immigrants entitled to vote, that is about 11,000. Eight candidates were registered, of different countries of origin: Philippines, Sri Lanka, Iran, Nigeria, Tunisia, Cape Verde, Ivory Coast and Eritrea. The first five were elected. A Tamil representative was elected for refugees. Surprisingly, some large communities, such as the Moroccan and the Mauritian, did not nominate any candidate, even if widely present in Palermo; also their participation in the election was particularly low.

This aspect needs further consideration and has to be fully considered by the Municipality in the near future. In the face of indifference and apathy, participation cannot be imposed by authority, nor can it be realized in a centralized form, but must necessarily develop within the groups and local communities themselves, giving rise to what could be defined as a collective process of self-legitimization (Friedmann, 1992a). Briefly informing and consulting the local community is not synonymous with participation. To have an effective (and efficacious) participation, a redistribution of power and the possibility to choose are necessary (Arnstein, 1969; Friedmann, 1992b; RTPI, 1983). By inviting ethnic communities to public hearings, a first sketch of fruitful information would be provided as well as a propitious climate for the Municipality's future work; it would also contribute to strengthening immigrants' presence in the political community (Friedmann and Lehre, 1997; Sandercock, 1998). One of the primary objectives is therefore to fight indifference within the local community, eliminating distrust in those who in the past have been excluded from decision-making processes, as it happened, for example, in the case of the multipurpose social center "Al-Khalisa". A platform of political objectives – along the lines of those expressed in the Municipality program – is necessary towards this goal, to win consensus and to direct conflict within the community towards common objectives.

Some members of ethnic communities pointed out that timetables and deadlines were quite short, and that there was little time to brief people and organize and support candidates. A Tunisian candidate (who then was not enabled to run for the election in consequence of his double Italian/Tunisian nationality) expressed doubts on the real value and significance of the initiative, reckoning that most of candidates were communities' representatives "close", in one way or another, to the administration.

As this first experience of elections shows, the legitimacy of representation, or better, of the faithfulness of individual representation in expressing the needs, desires and expectations of the community or of the group, remains a problem which is, in large part, unresolved. Once again, if this is a problematic aspect, independently of the plurality of the ethnic group in question, then a further degree of complexity can be ascribed to the presence of ethnic plurality, since there is less mediation using traditional forms of political representation (particularly with reference to the objective of involving the minorities themselves).

The elected representatives' office will last 24 months. Awaiting the right to vote for the immigrants and the institution of the City Consultative Assembly, the powers of the body elected will be to represent, give its advice and makes its proposal to the City Administration.

This election was strongly wanted and organized by Mayor Orlando, who firmly decided to go on with his policy of multiculturalism and integration, despite the general lack of support expressed by the City Council. In fact, for months the City Council had delayed the establishment of the City Consultative Assembly of Immigrants, and so the Mayor took the decision of calling the consultative election of some immigrants' representatives. "The Council has the right to fix its agenda and priorities, and I have mine", the Mayor said.

This election can also be viewed as something more than a symbolic gesture, and it represents a firm, significant political choice that stands out from national trends: at the local scale, it reaches goals and pays attention to immigrants' rights that have not been considered by national government up to now. In fact, the Parliament has not yet approved the bill enabling immigrants to vote for local government elections. On the other hand, the Mayor appears to be quite aware of the potential and implications of local initiatives: as a matter of fact on the same day of the election he wrote to the President of the Republic, the President of the Council of Ministers and the Presidents of the two houses of Parliament, informing them of the Palermo experience and results, and asking them for a meeting with the elected representatives.

What Space for Action? Possible Interventions for the Future

We have described above relatively new developments in Palermo, the presence and needs of ethnic minorities and their problems related to their living conditions in the historic center. In terms of urban dynamics and the uses and transformations of city spaces, the implications are many and problematic.

Despite the limited number of initiatives effectively implemented and the gap between intentions/programs and results, the policies for immigrants undertaken by the Palermo City Council can indicate a possible alternative by comparison with models of action widely adopted in the rest of Italy and in many other European cities. Risks are still present in the Palermo context, and those policies effectively implemented up to now do not appear sufficient: other strategies have to be explored, developing the undoubtedly positive directions and programs previously formulated.

Within the above-outlined framework, there are various specific contexts and structures which, for some aspects, can hardly be compared to other European urban realities. As a matter of fact, in Palermo the scenario is: not well-established migratory processes, (ethnic) plurality of resident immigrants, seemingly low level of external conflict versus a strikingly high level of internal conflict and/or disintegration, low level of maturity and awareness (as per goals and strategies) of local communities, a scant number of institutions committed, the key role of the municipal administration, multiple initiatives and/or outcomes with a low rate of interaction, poor level of expertise of local planners who are faced with a brand new reality characterised by new events and problems.

As an example, just think of "diversity" – compared to the standards commonly adopted – of a large part of the needs expressed. These needs relate to the specific character of the ethnic group resident in the area, and an outsider would hardly be able to identify them. The "anomalous" percentage distribution of the different age groups, the particular structure of a large part of the families, the "contradictory" role of women and the cultural and religious traditions all emerge as significant factors.

The comparison of family types and their distribution over the various districts of the city highlight once again the singularity of the historic centre[16] as regards both single-member families and single-parent families (21.5 percent) and the high number of "enlarged" families (9.5 percent) as well. Two main factors are at the root of this scenario: the presence of immigrants' "extensive" families (there are 10 foreign families out of 100 in the historic center, whereas the rate is one out of 100 in the rest of the city) and the forms of cohabitation of "enlarged" families belonging to (Italian) low-income classes subject to manifest social marginality and poverty (we should not overlook the fact that some areas of the historic center are still among the most marginal and socially depressed of the whole city).

A further analysis of the structure of the families (both Italian and foreign) living in the historic centre proves to be particularly useful in order to draw an outline of the general needs and requirements in public services, being quite different from the structure of families living elsewhere in the city. Recent surveys show that 10 percent of the families have two members, one of which is a minor, while the figure is five percent in the rest of the city; at the same time 29 foreign families out of 100 belong to this very family type, while only five percent of Italian families do. Moreover, since the migratory phenomenon is quite recent and the Italian population is undergoing ageing, the gap between Italians and foreigners is striking as regards elderly families with at least two members: 7.9 percent among Italians and not even 0.1 percent among foreigners (Attanasio e Giambalvo, 2001).

However, when compared to other urban realities, demands in Palermo are more incomplete, I would say less "mature" and unfit to be met with specific and sound projects: this is mainly due to the high number of ethnic groups, the high rate of marginality, low rate of cohesion and so far insufficient levels of political organization and self-determination. The character of recent immigration, shared by the majority of the communities in Palermo, is undoubtedly one of the main causes of this scenario. Nevertheless, particular emphasis should be put upon the role played by immigrants' communities in undertaking, for various reasons and in non-conventional forms, "conservative" actions for the sake of certain areas of the historic structure of the city, thus also guaranteeing a wealth of custom, benefits, social make-up and lifestyles which do not survive where processes of urban renovation and gentrification take place (see e.g., Winnick, 1990; Smith, 1991; Muller, 1993 and Carmon, 1998).

If it is acknowledged that the historic center should regain its previous residential use, the re-population will inevitably involve either the return of the middle and lower classes living on the outskirts or the retention of the present ethnic communities, trying to avoid social, racial and functional polarization (Cannarozzo, 1996 and 1999; Lo Piccolo, 1996; Caudo and Lo Piccolo, 1998). This process will not have any chance of getting started if the City Council is not able to organize effective housing and social policies, and to launch co-ordinated programs for safeguarding and reusing its estate so as to encourage private investment and to support the most disadvantaged groups.

In spite of all the remarkable differences one can spot from case to case, the above outline has however underlined the importance of redefining the role of local communities, and of ethnic groups in particular (Bourne, 1993; Carmon, 1998), within the framework of planning processes, starting from the relationship between planning demand and supply, and between forms of representation and participation.

In the Palermo case, consultation has turned out to be a positive means, whatever the results will be, to increase the expression and communicative potential of local communities as well as their awareness of representing an autonomous and active political subject, thus broadening and improving the democratic process. At the same time, consultations may turn out to be the key to legitimate already taken decisions, thus guaranteeing to local ethnic groups nothing more than what was previously and elsewhere decided to grant. In this view, any form of participation may be seen as a means of "pacification" in highly conflictual contexts, where any settlement of conflicts is not conditional upon a real and substantial change of original discrimination conditions.

Nevertheless it is maintained that through the very involvement of minority immigrants in the acts of transformation and governing a city it is possible to guarantee – beyond the traditional forms of political representation – a recognition of the rights to citizenship which are often denied on a political level, and almost always on a practical level. In other words the hypothesis which could emerge from the experience of the historic center of Palermo is one of participation in the planning process as an "alternative" form of representation on the part of the minorities discriminated against or excluded, or anyway unable to gain access to the traditional forms of representation: in other words, a form of insurgent citizenship (Holston, 1995; Sandercock, 1998).

Yet, as far as the Palermo case is concerned, these problems still have not been drawn into a broader scope of planning policies and so the matter needs further investigation both in the technical-disciplinary and political-administrative sectors. There are both conceptual and terminological factors, which give rise to preconceptions and misunderstandings. Significant examples are the uncertainties shown by the Palermo municipal planners in coping with the problem of standards for places of worship in relation to the increasing number of religions or the discriminatory effects of the regulation regarding funding for the refurbishment of the old housing estate.

However, in Palermo there are some areas of intervention in which this involvement can be experimented with, as the re-use of some historic buildings owned by the Municipality as places of worship for different faiths and pivotal public housing projects. In the housing sector the involvement of immigrants and minority communities could provide direct support to the implementation of recovery projects, by means of traditional low-cost techniques, by the adjustment of historic typologies to new and different needs and by sorting out, implementing and managing the related essential social services.

Palermo's unique model of action involves aggressive governmental leadership working in a coalition with non-profit cultural, educational, social and voluntary institutions. In recent years the city government, thanks to the leadership of an aggressive mayor, has started to map out a strategy to promote multiculturalism and integration. Attempts, success and failures experienced in Palermo highlight the need to revise the narrowness and the rigid character of the existing planning policies and practices. Furthermore, some positive aspects of the program run by the Municipality can be considered as an interesting and unconventional effort to guarantee those rights of citizenship, which are mostly denied on a national political level.

In many European cities the presence of immigrants in inner urban areas is very often seen as a social problem and a possible source of conflicts (Khakee et al, 1999). Even though the recent character and the limited range of the immigration phenomenon in Italy do not affect urban renewal processes as much as in other European countries, in the next few years the Italian scenario will progressively become similar to the Western European one in terms of presence of ethnic groups and problems of integration (Brusa, 1997). In this respect, many Italian cities are following European trends and models of action, a prospect that does not suggest good results and solutions of conflicts (Favaro and Tognetti Bordogna, 1989). The Palermo experience could be taken as the possible alternative for the future, even with all its unresolved problems and contradictory aspects.

Notes

1 The statistic, according to which Italy has the highest percentage of non-EC immigrants of its total number of foreigners, while presenting one of the lowest percentages of immigrants with respect to the national population, is highly significant.

2 The quantitative analysis is faced with several hindrances. ISTAT, Questura and the register office issue the most reliable data. ISTAT data only cover those foreigners who obtained a lawful residence permit in compliance with law 943/1986 but do not indicate actual residence of people.

3 From a general point of view, Palermo behaves like other big cities having a high number of ethnic groups, versus smaller cities, which tend to "specialize" in drawing some nationalities rather than others. As a consequence, if on a regional level about 80 percent of non-EC immigrants come from Central and Northern Africa, in Palermo the figure is lower than 60 percent.

4 Within a general context of urban decay, the historic center of Palermo has – like many other Southern Italian towns – specific characteristics and problems: a rich cultural

history; a great concentration of monumental and historic buildings; a complex and dynamic physical structure; increasing density of development; the physical decay of buildings; social deprivation of the inhabitants; depopulation; and slowness and inadequacy of public policies (Cannarozzo, 1990; Lo Piccolo, 1996; Caudo and Lo Piccolo, 1998).

5   On-the-spot observation and analysis of territorial distribution of the data issued by the register office was carried out by senior students Francesca Amenta, Ninni Barbagallo and Marisa Mirone.

6   The widest number of immigrants from Sri Lanka (368 men and 218 women, summing up to 586 out of a total of 3,448 in the city) is recorded in Mandamento Monti di Pietà (197 residents) and Mandamento Castellammare (232 residents). These people live in highly crowded dwellings and dislike cohabitation with other ethnic groups apart from those coming from Bangladesh since they share similar cultural heritage and traditions.

7   In 1993 the City Council required a survey to identify the characteristics and size of the special demand for housing (Gruttadauria, 1994). This survey was implemented in compliance with the guidelines and goals of the national housing law 179/92; in particular art. 4 of the law assigns a part of housing financial resources and initiatives to "particular social categories", that is to special needs groups such as the elderly, disabled people, former drug addicts, non resident students, immigrants. The law also recognizes that these special housing needs can be satisfied by standard housing units as well as by specific housing typologies, such as mini-units, communal housing, dormitories, hostels, accessible and sheltered housing (Gruttadauria, 1994, p. 6).

8   During the 1996/97 school year, 60 schools and about 100 teachers participated. Under the supervision of project personnel, sometimes immigrants themselves, the teachers contributed to the creation of an attitude to brotherhood, starting with stories dealing with their lives and experiences. The issues concerned some universal aspects of human experience transcending cultural differences: space, time, body, language, life's stages and the ways knowledge is disseminated.

9   It is important to remark that three years before national law 40 of 1998 on immigration was first passed, the Palermo municipal administration received a circular letter from former mayor Leoluca Orlando who, in compliance with European regulations, required from the city council the issue of by-laws including non-EC immigrants among those being eligible for social and healthcare support by the municipality. The only exception was the maternity allowance, which can only be granted to Italian citizens.

10   This is the case for loans to non-EC citizens. In 1998 the municipal administration arranged an agreement with S. Angelo Bank for granting loans to non-EC citizens for the purchase of durable goods. According to the agreement, the municipal administration commits itself to create a dedicated budget item and renew year by year the agreement with other banks. Only one single member of a given family may be the recipient of the one-year grant. Over the year 1998, more than two hundred foreign citizens benefited from the grant; unfortunately from 1999 the agreement was stopped. The bank, however, continues to provide non-EC citizens with some services, such as forms in several languages (Arab, English, French), currency change, credit/loans and saving accounts.

11   The goals of the project are: control over school attendance, job and professional training, development of educational, training and socialization patterns. The activities are carried out by volunteers, and include literacy and Italian language syllabuses, films in mother tongue, English and computer courses and after-school activities, which take place in the headquarters of the center, Via Discesa dei Giudici, in the historic center. Sports activities are carried out at the gym center of Via Orso Corvino and the S. Basilio Monastery.

12   At the beginning, activities were aimed at only 30 children, while today over 250 minors, from all over the world, but mostly from Bangladesh and Sri Lanka, attend the center.

13 The buildings will be united into one single building, thus creating 7 flats. The ground floor has: a waiting lounge, reception, video lounge, laundry and computer-equipped areas. The project provides for 7 lodgings including one for disabled persons.
14 While waiting for the law both granting non-EC citizens' rights to attend local government elections and establishing the municipal consultative assembly, the advisory council will have power of agency, consultation and proposal at the municipal administration of Palermo.
15 The nominees who were not backed by 30 valid supporting signatures have not been considered as valid. Every signer can only support one nomination, otherwise any support-signature will be made null and void.
16 As an example, we remind the reader that the historic center is an anomalous case also as regards the average number of members of families: such number is as low as 2.3 members while elsewhere over the municipality territory it is 2.9.

References

Arnstein, S. R. (1969), "A ladder of citizen participation", *Journal of the American Institute of Planners*, 35(4), pp. 216–224.

Attanasio, M. and Giambalvo, O. (2001), "Uno studio delle tipologie familiari nella città di Palermo", paper presented at *Giornate di studio sulla Popolazione*, Milan 20–22 February (mimeo).

Attili, G. (2001), "Nomadismo e sedentarietà: epistemologia visionaria per una nuova etica dell'abitare", in Scandurra, E. et al (ed), *Labirinti della città contemporanea*, Roma: Meltemi, pp. 57–80.

Barthes, R. (1981), "Semiology and the urban", in Gottdiener, M. and Lagopoulos, A. P. (eds), *The city and the sign: An introduction to urban semiotics*, New York: Columbia University Press, pp. 87–98.

Bird, J. (1992), *Alfredo Jaar: Two or Three Things I Imagine about Them*, London: Whitechapel Art Gallery.

Bourne, L. S. (1993), "The myth and reality of gentrification: a commentary on emerging urban forms", *Urban Studies*, 30, pp. 183–189.

Brusa, C. (ed.) (1997), *Immigrazione e multicultura nell'Italia di oggi*, Milano: Franco Angeli.

Cammarata, G., Meli, B. and Villa, E. (eds) (1996), *Many Peoples One City: Palermo. A Project for Immigrants*, Palermo: Municipality of Palermo – Committee for Social and Health Activities and Immigration Issues.

Cannarozzo, T. (1990), "Palermo Centro Storico", *Recuperare*, 48, pp. 338–349.

Cannarozzo, T. (1996), *Palermo tra memoria e futuro. Riqualificazione e recupero del centro storico*, Palermo: Publisicula.

Cannarozzo, T. (ed.) (1999), *Dal recupero del patrimonio edilizio alla riqualificazione dei centri storici. Pensiero e azione dell'Associazione Nazionale Centri Storici-Artistici in Sicilia*, Palermo: Publisicula.

Cannarozzo, T. (2000), "Palermo: mezzo secolo di trasformazioni", *Archivio di studi urbani e regionali*, 67, pp. 101–139.

Caritas diocesana di Roma (1994), *Immigrazione. Dossier Statistico*, Roma: Anterem.

Carmon, N. (1998), "Immigrants as Carriers of Urban Regeneration: International Evidence and an Israeli Case Study", *International Planning Studies*, 3 (2), pp. 207–225.

Castles, S. (1989), *Migrant Workers and the Transformation of Western Societies*, Ithaca: Cornell University Press.

Castles, S. et al (1984), *Here for Good: Western Europe's New Ethnic Minorities*, London: Pluto Press.

Castles, S. and Kosack, G. (1973), *Immigrant Workers and Class Structure in Western Europe*, London: Oxford University Press.

Castles, S. and Miller, M.J. (1993), *The Age of Migration. International Population Movements in the Modern World*, London: Macmillan.

Caudo, G. and Lo Piccolo, F. (1998), "Palermo, l'area metropolitana e la città consolidata: dinamiche, piani e politiche", in Talia, M. (ed.), *L'urbanistica nelle città del sud. Processi insediativi e nuove politiche urbane nelle aree metropolitane*, Roma: Gangemi, pp. 261–304.

Collinson, S. (1993), *Europe and International Migration*, London: Pinter.

Cusumano, A. (1976), *Il ritorno infelice. I tunisini in Sicilia*, Palermo: Sellerio.

Di Liegro, L. and Pittau, F. (1990), *Il pianeta immigrazione*, Roma: Edizioni Dehoniane.

Famoso, N. (1999), "L'immigrazione in Sicilia tra integrazione e diffidenza", in Brusa, C. (ed.), *Immigrazione e multicultura nell'Italia di oggi*, vol. II, Milano: Franco Angeli, pp. 200–212.

Favaro, G. and Tognetti Bordogna, M. (1989), *Politiche sociali ed immigrati stranieri*, Roma: La Nuova Italia Scientifica.

Fincher, R. and Jacobs, J. M. (eds) (1998), *Cities of Difference*, New York: The Guilford Press.

Fisher, R. and Kling, J. (eds) (1993), *Mobilizing the Community. Local Politics in the Era of the Global City*, Newbury Park and London: Sage.

Friedmann, J. (1992a), *Empowerment. The Politics of Alternative Development*, Cambridge, Mass. and Oxford, UK: Blackwell.

Friedmann, J. (1992b), "Feminist and Planning Theories: The Epistemological Connection", *Planning Theory*, 7/8, pp. 40–43.

Friedmann, J. (1995), "Migrants, Civil Society and the New Europe: The Challenge for Planners", *European Planning Studies*, 3(3), pp. 275–285.

Friedmann, J. and Lehrer, U.A. (1997), "Urban Policy Responses to Foreign Immigration: The Case of Frankfurt-am Main", *Journal of the American Planning Association*, 63(1), pp. 61–78.

Giacomarra, M. (1994), *Immigrati e Minoranze. Percorsi di integrazione sociale in Sicilia*, Palermo: La Zisa.

Granata, E. (1998), "Milano: città di frammenti", *Urbanistica*, 111, p. 34.

Gruttadauria, D. (1994), *La domanda marginale nel sistema residenziale di Palermo (mimeo)*, Palermo: Municipio di Palermo – Assessorato al Territorio – Ripartizione Urbanistica.

Healey, P. (1997), *Collaborative Planning. Shaping Places in Fragmented Societies*, London: Macmillan.

Holston, J. (1995), "Spaces of Insurgent Citizenship", *Planning Theory*, 13, pp. 35–51.

Indovina, F. (1991), "Segregazione etnica e strumentazione urbanistica", in Somma, P. (ed.), *Spazio e razzismo*, Milano: Franco Angeli, pp. 7–13.

Istituto Nazionale di Statistica (1993), *La presenza straniera in Italia. Una prima analisi dei dati censuari*, Roma: Istat.

Judd, D. and Parkinson, M. (eds) (1990), *Leadership and urban regeneration: cities in North America and Europe*, London: Sage.

Khakee, A. et al (ed.) (1999), *Urban Renewal, Ethnicity and Social Exclusion in Europe*, Aldershot: Ashgate.

Lanzani, A. (1998), "Modelli insediativi, forme di coabitazione e mutamento dei luoghi urbani", *Urbanistica*, 111, pp. 32–39.

Lazzarini, G. (1993), *La società multietnica*, Milano: Franco Angeli.

Loomba, A. (1998), *Colonialism/postcolonialism*, London and New York: Routledge.

Lo Piccolo, F. (ed.) (1995), *Identità urbana. Materiali per un dibattito*, Roma: Gangemi.

Lo Piccolo, F. (1996), "Urban Renewal in the Historic Center of Palermo", *Planning Practice and Research*, 11(2), pp. 217–225.

Lo Piccolo, F. (1999), "Ex Partibus Infidelium: Participation and Solution to Conflicts in the Experience of the Bengali Community in the East End of London", *Plurimondi. An International Forum for Research and Debate on Human Settlements*, 2, pp. 83–104.

Lo Piccolo, F. (2000a), "Città delle differenze, città dei diritti. Minoranze etniche, eque opportunità e conflitti urbani: considerazioni su un tema complesso", in Piroddi, E. et al (ed.), *I futuri della città. Mutamenti, nuovi soggetti e progetti, Milano: Franco Angeli*, pp. 411–445.

Lo Piccolo, F. (2000b), "Palermo, a City in Transition: Saint Benedict 'The Moor' versus Saint Rosalia", *International Planning Studies*, 5 (1), pp. 87–115.

Manconi, L. (1990), "Razzismo interno, razzismo esterno e strategia del chi c'è c'è", in Balbo, L. and Manconi, L. (eds), *I razzismi possibili, Milano: Feltrinelli*, pp. 45–91.

Manconi, L. (1992), "Luoghi e norme", in Mauri, L. and Micheli, G. A. (eds), *Le regole del gioco. Diritti di cittadinanza e immigrazione straniera*, Milano: Franco Angeli, pp. 99–107.

Massey, D. (1991), "The political place of locality studies", *Environment and Planning A*, 23, pp. 267–281.

McDowell, L. and Massey, D. (1984), "A woman's place?", in Massey, D. and Allen, J. (eds), *Geography matters! A reader*, Cambridge: Cambridge University Press.

Melotti, U. (1988), *Dal terzo mondo in Italia*, Milano: Centro studi Terzo Mondo.

Melotti, U. (1993), "Migrazioni internazionali e integrazione sociale: il caso italiano e le esperienze europee", in Pinna, M. (ed.), *L'Europa delle diversità. Identità e culture alle soglie del terzo millennio*, Milano: Franco Angeli, pp. 323–365.

Miller, M.J. (1981), *Foreign Workers in Western Europe: an Emerging Political Force*, New York: Praeger.

Muller, T. (1993), *Immigrants and the American City*, New York: New York University Press.

Paba, G. (1998), "Cortei neri e colorati: itinerari e problemi delle cittadinanze emergenti", *Urbanistica*, 111, pp. 20–24.

Pacini, M. (1989), "Transizione demografica, migrazioni internazionali e dinamiche culturali", in Fondazione Giovanni Agnelli (ed.), *Abitare il pianeta. Futuro demografico, migrazioni e tensioni etniche – Volume Primo. Il Mondo Arabo, l'Italia e l'Europa*, Torino: Edizioni della Fondazione Giovanni Agnelli, pp. 3–45.

Perrone, C. (2000), "Firenze Insurgent City: Atlante delle nuove pratiche sociali urbane", paper presented at the workshop *Prove di realizzazione di atlanti territoriali*, Gavorrano (Grosseto) 20–21 October (mimeo).

Pinna, M. (ed.) (1993), *L'Europa delle diversità. Identità e culture alle soglie del terzo millennio*, Milano: Franco Angeli.

Prashar, U. and Nicholas, S. (1986), *Routes or Roadblocks? Consulting minority communities in London boroughs*, London: The Runnymede Trust, The Greater London Council.

Ratcliffe, P. (1998), "Planning for Diversity and Change: Implications of a Polyethnic Society", *Planning Practice and Research*, 13(4), pp. 359–369.

Royal Town Planning Institute/Commission for Racial Equality Working Party, (1983), *Planning for a Multi-Racial Britain*, London: Commission for Racial Equality.

Sandercock, L. (1998), *Towards Cosmopolis. Planning for Multicultural Cities*, Chichester: John Wiley and Sons.

Sandercock, L. (2000), "When Strangers Become Neighbours: Managing Cities of Difference", *Planning Theory and Practice*, 1(1), pp. 13–30.

Scandurra, E. (1999), *La città che non c'è. La pianificazione al tramonto*, Bari: Edizioni Dedalo.

Scandurra, E. (2001), *Gli storni e l'urbanista. Progettare nella contemporaneità*, Roma: Meltemi.

Sennett, R. (1990), *The conscience of the eye: The design and social life of cities*, New York: Knopf.

Sibley, D. (1995), *Geographies of Exclusion. Society and Difference in the West*, London and New York: Routledge.

Smith, M.P. et al (1991), "Coloring California: new Asian immigrants' households, social network and local state", *International Journal of Urban and Regional Research*, 15(2), pp. 250–268.

Soja, E. W. (1989), *Postmodern Geographies. The Reassertion of Space in Critical Social Theory*, London and New York: Verso.

Somma, P. (1999), "Ethnic minorities, urban renewal, and social exclusion in Italy", in Khakee, A. et al (ed.), *Urban Renewal, Ethnicity and Social Exclusion in Europe*, Aldershot: Ashgate, pp. 73–104.

Thomas, H. (2000), *Race and Planning. The UK Experience*, London: UCL Press.

Thomas, H. et al (1996), "Locality, urban governance and contested meanings of place", *Area*, 28 (2), pp. 186–198.

Tosco, M. (1994), "Extracomunitari a Palermo", *Urbanistica Informazioni*, 138, pp. 62–64.

Tosi, A. (1998), "Lo spazio urbano dell'immigrazione. Una problematica urbana", *Urbanistica*, 111, pp. 7–19.

Winnick, L. (1990), *New People in Old Neighborhoods*, New York: Russell Sage Foundation.

Acknowledgements

The author would like to thank Ornella Giambalvo, associate professor at the School of Economics of the University of Palermo, for the irreplaceable help and guidance on statistical data, and Vincenzo Genna, CISS operator, for providing detailed information on local initiatives for immigrants undertaken in both the voluntary and institutional sectors. Ideas, comments and field research were shared with my students Francesca Amenta, Ninni Barbagallo and Marisa Mirone: to them many thanks for the opportunity to benefit from such an enriching experience. Special thanks to Roberto Lanzi, who generously gave a great help in the English version of this text, thus being "diverted" from his main work and other things. Last but not least, many thanks to Huw Thomas for his usual support, criticism and patience.

Chapter 7

On Our Terms: Ethnic Minorities and Neighborhood Development in Two Swedish Housing Districts

ABDUL KHAKEE AND BJÖRN KULLANDER

Introduction

Interest organizations have played a considerable role in the development of the Swedish welfare state. They have participated in public decision-making at all levels of government. Since the 1980s such participation has declined appreciably at the national level (Lewin, 1994; Rothstein and Bergström, 1999). However, this is not so at the local government level where it is quite substantial.[1] One such organization consists of tenants' associations that have played an important part – both nationally and locally – in the development of the housing sector in which the public control has been extensive during the entire period of the evolution of the welfare state.

In Sweden the public ownership is equivalent to about half of the total stock of rental houses. A substantial portion of public rental housing was built as a part of the Million Program (between 1965–74 Sweden built one million apartments, mostly in the suburban areas of large cities and towns in order to house the increasing urban population). As opposed to other western countries, immigrants to Sweden have not been able to find housing in inner-city areas because of high rents and considerable tenant-ownership (Murdie and Borgegård, 1998). Successive waves of immigrants (labor migrants during the 1960s and early 1970s and refugees since the mid-1970s) have been allotted housing in the Million Program districts in major cities and towns. It is these districts that have become the target of the public debate and academic research on social segregation and urban regeneration (see e.g. Andersson, 1997, 1998; Andersson-Brolin, 1984; Andersson and Molina, 1996; Biterman, 1993, Lindén and Lindberg, 1991; Molina, 1997, Westin and Dingu-Kyklund, 1997).

The issue at hand is to what extent have the immigrant groups or ethnic minorities[2] living in the Million Program districts been able to affect policies in order to counteract social segregation and promote urban regeneration. The chapter presents two case studies. The first case is based on an attitude survey among Swedish and immigrant members of tenants' associations in the district of Brickebacken in the city of Örebro. It presents an account of the attitude of the Swedish and ethnic minorities towards participation and representation. The second case describes the role of immigrant groups in urban regeneration in the district of Rinkeby in Stockholm. It discusses how dissatisfaction with property-led urban renewal stimulated immigrant groups takes a more active part in the following

Knights and Castles

program for urban regeneration. Prior to the presentation of the case studies, the authors look at the specific aspects of housing segregation and ethnic minorities' representation in interest organizations. The concluding section presents some reflections on the complicated nature of social segregation, ethnic participation and urban regeneration.

Housing Segregation: Swedish Characteristics

As mentioned above, issues related to housing segregation in Sweden are fairly well documented (see references cited above). The aim of this brief section is only to highlight some of the major differences between the situation in Sweden in relation to other western countries.

Housing segregation in Sweden is a *suburban problem* rather than an inner city one. A recent OECD survey shows that the proportion of ethnic minorities living in poorer suburban areas in Sweden is higher than in other member countries (OECD, 1996). These poorer areas are generally the Million Program housing districts that from the very beginning got poor publicity. These are high rise and high-density suburbs built with mass production techniques. The physical environment outside the buildings is indigent and poor. These suburban districts have got the stigma of "problem areas" characterized by vandalism, violence, criminality, drugs and alcoholism (Goldfield, 1979). It is in these areas that the majority of the immigrant households live.

The Swedish housing market has a special structure that acts as a social and ethnic sorting-out instrument. It is composed of three major categories: rental apartments – 42 percent of the housing stock; tenant-owned and co-operative flats – 18 percent; and owner occupied houses – 40 percent. Nearly half of the rental apartments belong to public housing. A substantial stock of the rental housing in the Million Program areas belongs to the latter category.

A study of the differences in housing occupancy for Swedish and immigrant households respectively is shown in Table 7.1.

The figures in the table show immigrant households are over-represented in rental housing, most of which is public housing. Two thirds of Swedish households live in their own houses or flats. The corresponding figures for immigrant households is 50 percent for households who came to Sweden before 1975 and often had paid employment on arrival and 23 percent for immigrant households who came after 1975. Several reasons are put forward to explain the poor "housing career" among the newly arrived immigrant households:

* socio-economic status (especially lack of capital or other form of security for purchasing an apartment or a house);
* length of residence in Sweden;
* discriminatory practices of private landlords;
* local housing authorities' policy of systematically directing immigrant families to areas with high concentration of immigrant households (Molina, 1997).

The impact of "push" and "pull" factors accounting for segregation to specific rental housing areas has been a controversial issue. Some argue that attempts to reduce or eliminate segregation may counteract the attempts towards multiculturalism. In fact, many immigrant groups prefer to live together (see, e.g. Borgegård, 1996). Others claim that the contention that immigrant households prefer to live near their compatriots or other immigrant households is nothing but cultural reductionism rhetoric (Molina, 1997). A factor, specific for Sweden that has lessened ethnic minorities' ability to do a "housing career" is the "whole-of-Sweden" policy encouraging all local governments to take a fair share of immigrants and refugees. Many of these households move to Stockholm or other larger cities after their refugee reception period only to start at the bottom of the market in order to find suitable housing (Murdie and Borgegård, 1998).

Housing segregation in Sweden, as in Britain, has been increasingly "race selective" (see e.g. Andersson, 1996; Robinson, 1987). Certain minorities are discriminated against more than others. Cultural differences between ethnic minorities and the majority population constitute an important factor determining the level of segregation. The greater these differences the greater is the level of segregation. A recent calculation of a segregation index[3] for some ethnic minorities in 14 Swedish municipalities are shown in Table 7.2. Table 7.2 confirms other studies that show that groups with smaller cultural distance to the majority population (e.g. Finns, Poles, Germans, Italians), exhibit lower levels of segregation whereas households from Asia Minor and Africa show the highest level (see, also, Murdie and Borgegård, 1998). Moreover, the level of segregation has increased as a result of the changing nature of immigration. Refugees from southeast Europe and Asia Minor dominate the immigration since 1980s. Table 7.2 also shows that the later immigrants are more segregated than the earlier ones, almost regardless of the country of origin. This confirms our previous observation, namely that a refugee household has economic as well as social disadvantages in finding a suitable home.

Public officials in Sweden contend that there is a "tipping-point" or a "critical size" of non-Nordic population in a housing district beyond which Swedes and other northern Europeans are "scared away". This is estimated to be about 25–30 percent (Wirtén, 1998). In a study of Uppsala, Sweden's fourth largest city, Molina (1997) shows that housing segregation is closely related to increasing racial tendencies in Swedish society. Appearance is decisive. The more non-Nordic an appearance the more is repudiation and discrimination. In this fashion the urban space in Sweden is increasingly radicalized (see also, Khakee and Johansson, 1999).

Social segregation is a comprehensive phenomenon extending beyond housing to encompass almost all domains of life. Ethnic households need to develop social and economic networks. Paid employment is a crucial factor in developing these networks. Discriminatory practices in the labor market seem to be more severe in Sweden than elsewhere in western countries. Requirements of being "Swedish" do not only mean good knowledge of the Swedish language but also include diffuse criteria termed "social competence". Many are the tales of immigrants with good education and knowledge of Swedish and other languages denied entrance into the labor market because of an odd surname! The discriminatory labor market practice hits the black population badly, for example, many adult-Somalis in Sweden lack regular employment. Legislation against discrimination in the labor market has so far not had any decisive impact.

Ethnic Minorities' Representation

In the Swedish debate on representation a distinction is made between "opinion representation" and "social representation". The former implies an agreement in the opinions of the electorate and the elected representatives whereas the latter refers to the presence of different social groups in the political and interest organizations. In Sweden there has been until quite recently strong support for opinion representation (see e.g. Rothstein, et al., 1995; Westerståhl and Johansson, 1981). Two factors that have been decisive in the shift from opinion to social representation are the development of Sweden towards a multi-ethnic society and the impact of feminist ideology. The latter has emphasized equal representation of women as a democratic right. Feminist arguments can be applied equally to ethnic minorities' representation. Besides the question of democratic right, the representation of ethnic minorities ensures the provision of knowledge and experience that complements the knowledge and experience of the majority representatives. Thirdly the majority and minorities have different interests and values. Only through a proportional representation of ethnic minorities can these differences be made perceptible. Hernes (1982) presents similar arguments with regards to women's representation.

Sweden is a highly organized society. This is reflected in the fact that 92.9 percent Swedes are members of an interest organization. The corresponding figure for ethnic minorities is 88.5 percent. On average each Swede is a member of three organizations, the corresponding figure for ethnic minorities is 2.3. 52.4 percent of Swedes are *active* in at least one organization with 29.9 percent having a commission of trust. The corresponding figures for the immigrants are 43.4 percent and 21.5 percent respectively (Häll, 1994). These are aggregate figures and do not show how the representation of various immigrant groups differs (interested readers are referred to Amin, 1996; Bäck, 1999; Bäck and Soininen, 1996; Khakee and Johansson, 1999; Rodrigo Blomqvist, 1997).

Interest organizations are classified into four major categories:

- work-related organizations, mainly trade unions;
- ideological organizations (excluding political parties) e.g. environmental, temperance and peace movements, religious societies, fraternal orders;
- identity organization e.g. housing associations, patients' organizations, women's movements, consumer co-operatives, ethnic associations;
- leisure organizations e.g. sports and cultural associations.

There is some overlapping between the four categories of organizations. The classification, nevertheless, presents a useful way of examining social participation.

Table 7.3 shows that every sixth Swede is active in the trade unions. The level of active participation for immigrants in trade unions is slightly less. This is so even when it comes to ideological and identity organizations. The only significant difference between Swedes and immigrants is in the active involvement in leisure organizations – more than three times as high for the Swedes compared to immigrants. With regards to membership in identity organizations Swedes' membership is higher than immigrants and as Häll (1994) points out the difference would be notable if membership in various immigrant associations is deducted (see also, Bäck and Möller, 1997). But the most significant aspect with regard to membership and the level of participation is the difference between various immigrant groups. In the absence of reliable statistics, we can only be speculating that minorities that are culturally different from the Swedes would be underrepresented in interest organizations. There is some evidence for this when we look at the figures for representation in the tenants' organization (see Table 7.4).

The tenants' association[4] is territorially organized with national, county, municipal and local tenants' associations. The organization has among its goals decent housing at reasonable cost, provision of auxiliary services and a good physical and social environment in the housing districts. The impact of tenants' associations on housing policy at national and local level has varied depending on political and economic circumstances.

Table 7.4 shows that a relatively high proportion of immigrants are members of the tenants' organization compared to other identity organizations (see Table 7.3). This is not surprising because the proportion of immigrant households who live in rental apartments is relatively high. However, Table 7.4 also shows that the membership in the organization is confined to households with Nordic or other European origin.

Participation and Representation in Urban Regeneration

Brickebacken: Some contextual facts

Brickebacken is one of the fourteen municipal districts in the municipality of Örebro. The municipality is the eighth largest in Sweden with a population of 122,641 inhabitants. 96,693 persons (80.5 percent) are of Swedish origin whereas 23,948 persons (19.5 percent) have immigrant background. About one-half of the latter (10,268 or 8.4 percent of the total municipal population) are second generation immigrants.[5]

Örebro has belonged to the country's industrial belt in central Sweden. It was formerly an important center for food and shoe manufacturing. Örebro lost many of the traditional industries during the structural transformation of the Swedish economy in the 1960s and 1970s. Industrial employment currently amounts to 21 percent of the gainfully employed, whereas 78 percent are employed in private and public services.

The housing district, Brickebacken, was built between 1969–73 as a part of the Million Program. Its 1800 apartments are located varyingly in six-, three- and two-storey buildings. The center of the district has commercial, cultural and social services. Brickebacken has the largest concentration of ethnic minorities in Örebro. Its 3,400 inhabitants are divided as follows:

- 55 percent are Swedes;
- 45 percent have immigrant background (38 percent of these are second generation immigrants). Less than one-fourth of the immigrants come from the Nordic and EU countries; one fourth come from the rest of Europe and just over half from the rest of the world. Among the latter, immigrants from the Asia Minor dominate.

Table 7.5 provides the employment situation for foreign-born and the entire population of working age in Örebro and Brickebacken respectively in 1996.

The figures show that the employment rate for the entire population in Brickebacken (and this applies to many other Million Program areas) is lower than the population living in other areas. Moreover, two thirds of immigrants in Brickebacken are unemployed. A fate they share with immigrants in other Million Program districts.

Brickebacken is divided into 14 housing neighborhoods, each with a tenants' association. These neighborhood associations together form the district tenants' association with one representative from each neighborhood association. The Brickebacken Association meets every month and is responsible for negotiating with representatives of Örebro Municipal Housing Company and also maintains contacts with politicians in the district council.

The Brickebacken Association has 921 members, of whom 190 (or 21 percent) are immigrants. Of the 80 persons with commission of trust, only 7 (or 9 percent) have an immigrant background. There is no representative of the ethnic minorities in the executive committee of the Brickebacken Tenants' Association.

Opinions on Participation and Representation

This section is based on interviews with nine persons in Brickebacken: three immigrants who are only members of the neighborhood tenants' association, three immigrants who have moreover commissions of trust in neighborhood associations' executive committee and three Swedes in similar positions. Among the latter, one person is a member of the election committee. All the six foreign-born interviewees were selected from non-European ethnic groups. The interviews dealt with questions regarding participation and representation.

The views of these groups, as they were expressed in the interviews, are presented in Tables 7.6a-c.

The interviews, summarized in the tables, give us both positive and negative aspects about participation and influence of ethnic minorities in the policy process related to housing and improvement of the housing district.

With regard to representation in the tenants association's executive committee, all three groups feel that immigrant representation is valuable, immigrants are underrepresented and that efforts should be made to increase the number of immigrants in executive positions. However, motives for increased immigrant representation differ. The Swedes have a more utilitarian view. They feel that an active participation on the part of immigrants would enable the latter to "learn" the norms and values of the Swedish society. It would make it easier to handle problems that are specific for various immigrant groups. The neighborhoods would function better if everyone followed the Swedish way of life. The interviewees from the ethnic minorities felt that their participation would enrich the policy process. They would come with proposals that are based on their life experience. It would also imply that the Swedes would be forced to take into account viewpoints other than those they are accustomed to.

Both the Swedish and non-Swedish representatives are in agreement that active participation in interest organizations is an essential factor in integrating minorities in Swedish society. Such participation generates social capital that plays a vital role in the political and social integration process (see, e.g. Putnam, 1993). It might be of interest to note that a recent national survey showed that about 17 percent of immigrants in Sweden were willing to work actively in interest organizations. However, these organizations tend to be conservative and are not inclined to change their ways of working in keeping with the new realities of a multicultural society (Micheletti, 1998).

Foucault has described how power-relations in a discourse are established through the language and concepts that are used. Discourses are used to exclude and draw up borders, distort facts and construct what can be regarded as true or false (Foucault, 1993). A closer examination of the interviews in Bricebacken shows that by describing immigrant groups as "the others" it is possible to establish certain perceptions about these minorities. "The others" are welcome to become members and participate actively; no one stops them. However, "the others" do not have the same appreciation of the "club culture", they think otherwise and they do not understand what are considered as acceptable norms and values. In fact the existence of formal rights for participation acts as an alibi for the majority representatives not to work actively for involving ethnic minorities. The interviews show that the Swedish representatives were aware that many persons from specific immigrant groups had previous experience of interest organizations from their native countries but it was easier to lump together the minorities and contend that immigrants generally lack appreciation of voluntary activities.

With regards to ethnic minorities' active participation in policy processes, minority representatives feel that there is limited scope for ethnic minorities to come with proposals. When such a scope exists, their views are not taken very seriously. Five major reasons for this are:

- lack of proficiency in the Swedish language that makes it difficult to articulate proposals in an "acceptable" fashion;
- level of unemployment among immigrants reducing their propensity to participate in civic activities;
- absence of support from a social network normally available to the Swedes;
- the prevalence of "we" and "the other" in the public discourse drawing up a border between Swedes and ethnic minorities;
- the procedure for getting nominated on the executive committee is exclusionary in nature; only those with a wide social network or those who are personally acquainted with the election committee have any chance in getting elected.

From Protest to Participation

Rinkeby: A Swedish Melting pot?

Rinkeby is one of the seventeen districts of Stockholm, the capital of Sweden. Stockholm has a population of three-quarter million inhabitants, of whom about 18 percent have an immigrant background.

Of Rinkeby's population of 14,900 inhabitants in 1998, 11,160 or 74.9 percent have an immigrant background. It is the most cosmopolitan housing district in the whole of Scandinavia, with over 100 languages spoken in the area. 99 percent of the school children do not have Swedish as their mother tongue though many of them are Swedish citizens and in fact born in Sweden. The ethnic composition of the population has changed during the past few years. Whereas the Finnish, Greek and (to some extent) Turkish population has declined, there has been rapid increase in the size of Somalis, Iraqis and Bosnians. The six largest immigrant groups are Turkish (12.3 percent), Somali (9.3 percent), Iraqi (5.8 percent), Greek (5.7 percent), Syrian (3.9 percent) and Finnish and Bosnian (3.7 percent each).

Like many other districts with a high immigrant population, Rinkeby exhibits a large number of characteristics indicating the vulnerable situation of the population: high rates of unemployment, a large share of the workforce in blue-collar jobs, low levels of income, greater proportion of households receiving social allowance, larger numbers of persons receiving early retirement pension and a lower percentage of over-19 population with post-upper-secondary school education (see further, Khakee and Johansson, 1999).

Rinkeby's 5,200 apartments are all in high rise buildings. 80 percent of them are managed by three public companies: Familjbostäder, Svenska Bostäder and Stockholmshem. The district has undergone two rounds of urban regeneration. The first one, implemented in the 1980s, emphasized physical improvement of the housing stock and the neighborhoods. The second is a more co-ordinated program initiated in the 1990s and is still in progress. There are substantial differences between the two programs with regards to the role of minorities in influencing urban regeneration.

Bay-window Renewal

Two of the three public housing companies in Rinkeby undertook physical improvement on a modest scale that on the whole was favorably received by the tenants (Sangregario, 1989). But the third company, Familjbostäder launched an extensive renewal that cost over 650 million Swedish Crowns. It included improving the standards and size of the flats – converting them into maisonettes with bay windows and clothing the grey concrete buildings in red bricks.

The plans for renewal led to strong protests and mobilized people in various contact committees. However, there was no unequivocal criticism of the plan. The major concern was that the program was expensive, that it would lead to higher rents and many of the tenants did not like the idea of moving twice – once to a temporary home and then back to the renovated apartment.

In fact Familjbostäder had in mind a change in the tenancy system by introducing selective elimination. Stable households, preferably Swedish, would be encouraged to rent the renovated flats and households with social problems would not be given apartments. At the advice of the Ombudsman against Ethnic Discrimination, the local authorities did not introduce such a regulation.

The result of the Familjbostäder's renewal program was not the one that the company had hoped for. The ethnic composition did not change significantly. Relatively well to do Swedish and immigrant households did not move back, which in fact increased ethnic concentration. The company, however, succeeded in debarring households living on social allowance, by denying such households a lease. A side effect, but a significant one, was the creation of a strong sense of belonging to Rinkeby among the inhabitants (see e.g. Ehn, 1992; Hanström and Johansson, 1994; Öresjö, 1996 and Sangregorio, 1989). The latter played a crucial role in the second round of urban regeneration.

Co ordinated Urban Regeneration

Both in Swedish and international literature it has been repeatedly pointed out that property-led and turn-around urban renewals do not lead to desired consequences. Besides the fact that they focus on physical measures and are costly, they are often based on the "beggar-your-neighbor" principle of exporting the negative impact to adjacent areas. They in fact do not solve the segregation problem (see e.g. Carlén and Cars, 1991; Power and Tunstall, 1995). Their top-down manner does not inculcate a sense of responsibility so vital in the conservation of neighborhood and for social integration.

The second round of urban regeneration in Rinkeby is based on principles of mobilization, participation and co-ordination of public and private efforts. Pride in the neighborhood, a sense of satisfaction with public and private amenities, responsibility for neighborhood environment, feeling of security and participation are touchstones of this regeneration (see, Cars and Edgren-Schori, 1999 for another similar Swedish effort).

The current Rinkeby project is a collaborative effort covering several policy sectors – housing, physical planning, labor market, education and social welfare and involving local government departments, other public agencies, civic associations and businesses. The objective is to show the need for a cumulative effort in order to attain substantial improvement in the socio-economic and physical condition of the neighborhoods. But the collective efforts also reflect the recognition of claims, concerns and interests that were articulated by the local population during the earlier renewal project and were subsequently voiced in other policy areas.

The project does not involve any significant renewal of the housing stock. The emphasis in this respect is the improvement of maintenance and upkeep of the neighborhoods. For example in some quarters a collective washhouse has been installed instead of a laundry room for each house. This has not only reduced wearing down of the washing machines but also improved the social ties between households belonging to different ethnic backgrounds.

With a relatively high unemployment level, a labor market program has been considered a crucial feature of urban regeneration. For this purpose, several unconventional methods have been adopted. A "House of Enterprise" has been established where premises are provided for a limited period of time for immigrants who want to start small businesses. It provides technical equipment, shared reception services, customer research and marketing services. Another project, "Integra" aims especially to provide trainee and apprenticeship program for immigrants with economic qualifications. This EU-supported project involves private businesses willing to provide trainee programs. A third project, "Wake Up" provides training for immigrant youths in various vocations including IT, filmmaking, service management, etc. The Rinkeby Employment Office has special grants from the European Union and the County Labor Board in order to activate long-term unemployed immigrants and improve their language and social competence. The labor market program is co-ordinated with the social insurance office's efforts to reduce people living on social insurance or early retirement pension (Rinkeby Stadsdelsförvaltning, 1999).

At the same time the district administration has launched an extensive educational program that includes the establishment of two folk high schools, upgrading secondary and post-secondary education in order to overcome handicap among children owing to bilingualism and a special educational program tailored for adult immigrants (ibid).

The co-ordinated regeneration program includes various cultural and social projects. SAFE-Rinkeby is a co-operative effort to reduce juvenile delinquency, drug abuse and various other crimes. Rinkeby Kulturskola is another establishment providing a broad range of cultural training programs.

All these efforts have contributed towards the creation of the sense of local multiethnic identity – "the other Rinkeby".[6] This is especially so with young people who with culture as a weapon try to earn respect and recognition. The district has become a seat of alternative culture with such nationally well known Hip-hop and antiracist singers as Lucco and Dagge who not only sing in protest against exclusion and neglect of human beings on account of their ethnic origin but also attempt to mobilize against drug abuse and truancy. These expressions forebode a new boundless multiethnic community (Åhlund, 1995).

Ethnic Minorities, Participation and Urban Regeneration

Urban regeneration in Sweden from the ethnic perspective is essentially a suburban issue. The major focus is on the Million Program housing districts where a majority of the immigrant population dwells. Brickebacken and Rinkeby are two such areas. The two cases provide somewhat different stories but they nevertheless provide useful insight into issues relating to social segregation, participation of ethnic minorities in civic associations especially those working for the improvement of neighborhood and proper approaches for urban regeneration.

The existence of housing areas with a concentration of ethnic minorities is a complex issue. Besides the fact that the presence of these areas cannot only be explained in terms of push factors, such areas are seen by some sociologists and ethnologists as seats of alternative culture that contribute towards the creative diversity of an urban region. Despite the persistence of social and economic problems and tensions between various immigrant groups residing in these areas, there is an increasing pride among young people in their neighborhoods. Moreover there is an increasing recognition that new forms of networking pave way to employment opportunities, social cohesion and local identity (Åhlund, 1995). The negative aspects of segregation can be curtailed by means of culturally sensitive planning that allows for exceptions and variances from planning policies and standards on a case-by-case basis (Qadeer, 1994). Such planning enhances the manifestations of multiculturalism in a city region and provides potentialities for development through cultural diversity.

The lack of participation of ethnic minorities in civic associations is not a simple issue about how these associations are organized and how members are recruited to their executive boards. Organizational set-up may be exclusionary in nature. Nevertheless other aspects are equally important. The social and economic circumstances act as barriers to participation. Lack of language proficiency, unemployment and dependence on social allowance act as constraints in active participation. However, it seems that despite these shortcomings, it is possible to establish meaningful dialogue between immigrant groups and the majority population. Better use of immigrants' own associations is one way to get immigrants involved in civic activities (see Holm and Khakee, 1994). The case of Rinkeby in this paper suggests the need for catalysts either in the shape of specific events or persons. Brickebacken's case suggests the need for removing linguistic and other barriers for a rational discourse between immigrants and the majority population. Both case studies provide useful insight into the need for networking in order to improve representation and to influence public policy processes.

The paper also suggests that a proper approach to regeneration is more than a question of physical improvement or the use of bottom-up methods involving various ethnic minorities. The way such regeneration is carried out is important. Nevertheless, the complex nature of regeneration implies that this is possible only if a co-ordinated outlook is applied. Regeneration is not only a physical process but also a social, educational and economic process requiring participation of a large number of stakeholders willing to develop various forms of networking and co-operation.

Notes

[1] At the local level interest organizations' role varies depending on the policy area (see, Gustafsson, 1993).

[2] Although the term 'immigrants' is generally used in the public debate and official statistics, it lacks a general definition. Usually immigrants include all persons who are registered for census purposes. These persons are foreign-born but they may or may not have retained their foreign citizenship. Moreover, households, where at least one of the parents is born abroad, are classified as 'immigrant households'. That means that statistics often refer to first and second generations of immigrants. In this chapter we shall use either the term 'immigrants' or 'ethnic minorities' to refer to both the first- and second-generation immigrants. (See, further, Khakee and Johansson, 1999).

[3] The segregation index is based on the idea of 'ethnic hierarchy' or the social position of various ethnic minorities. Andersson attempts to find out if the non-European immigrants

occupy a more subordinate position than the European ones with respect to residential location. 14 municipalities were chosen mainly on the basis of the size of the immigrant population and the index was calculated by comparing the distribution of the Swedish population and various immigrant groups.

4   Tenants' associations date from the First World War. They merged together into one national organization in 1923. Inter-war years were characterized by conflicts and confrontations between tenants and landlords – often resulting in mass blockades and evictions. Following the Social Democratic Party's assumption of power in the late 1930s, the tenants' organization was accorded a role in public policy making in the housing sector.

5   The composition of the immigrant population is as follows:
   - 5.8% come from other Nordic countries mainly Finland (3.9%) and Norway (1.2%);
   - 5.4% come from the rest of Europe (1.3% from Bosnia, 1.1% from Rest-Yugoslavia and 0.5% from Poland);
   - 1.8% come from Africa mainly from Ethiopia (0.4%) and Somalia (0.7%);
   - 5.8% from Asia divided between Iraq 0.8%, Iran 0.9%, Lebanon 0.8%, Syria 0.7% and Turkey 1.6%;
   - 1.1% from the rest of the world including 0.4% from Latin America, mainly Chile.

6   Rinkeby has in the media and public debate been synonymous with all that has gone wrong with a suburban ghetto. This is the common picture of the housing district. The other one that has been gradually evolving is shared by many of the inhabitants who find Rinkeby a place that offers warmth, excitement and cross-cultural contacts.

References

Åhlund, A. (1995), "Ungdomar, gränser och nya rörelser" ("Youth, boundaries and new movements"), *Rasismens varp och trasor (Racism's Warp and Rags)*. Public Report. Norrköping: Statens invandrarverk.

Amin, J. (1996), *Invandrarnas politiska representation (Immigrants' Political Representation)*. Department of Political Science, Umeå University, Umeå. (Master's thesis).

Andersson, R. (1997), "Divided cities as a policy-based notion in Sweden", Paper submitted at the *Nethur Conference on Undivided Cities*. The Hague.

Andersson, R. (1998), "Socio-spatial dynamics: Ethnic divisions of mobility and housing in post-Palme Sweden", *Urban Studies* 35(3) pp. 397–428.

Andersson, R. and Molina, I. (1996), "Etnisk boendesegregation i teori och praktik" ("Ethnic housing segregation in theory and practice"), *Vägar in i Sverige (Access to Swedish Society)*. Public Commission Report SOU 1996:55, Stockholm: Allmäna Förlaget.

Andersson-Brolin, L. (1984), *Etnisk bostadssegregation (Ethnic Housing Segregation)*. Stockholm: Swedish Council for Building Research.

Bäck, H. (1999), "Invandrarnas deltagande i det politiska livet" ("Immigrants participation in political life"), *Invandrarskap och medborgarskap (Immigrant Status and Citizenship)*. Public Commission Report SOU 1999:8, Stockholm: Fakta Info.

Bäck, H. and Möller, T. (1997), *Partier och organisatioer (Parties and Interest Organisations)*. Stockholm: Publica.

Bäck, H. and Soininen, M. (1996), *Invandrarna, demokratin och samhället (Immigrants, Democracy and Society)*. Göteborg: School of Public Administration, Göteborg University.

Biterman, D. (1996), "Den etniska boendesegregationens utveckling i Stockholms län 1970–1993" ("The ethnic housing segregation in Stockholm County 1970–1993"), in Bohm, K. and Khakee, A. (eds) *Etnicitet, segregation och kommunal planering (Ethnicity, Segregation and Planning)*. Stockholm: Nordplan.

Carlén, G. and Cars, G. (1991), "Renewal of large-scale post-war housing estates in Sweden: Effects and efficiency" in Alterman, R. and Cars, G. (eds) *Neighborhood Regeneration: An International Evaluation*. London: Macmillan.

Cars, G. and Edgren-Schori, M. (1999), "Social integration and exclusion: The response of Swedish society" in Allen, J., Cars, G. and Madanipour, A. (eds) *Social Exclusion in European Cities: Processes, Experiences and Responses*. London: Jessica Kingsley.

Daun, Å. (1996), *Swedish Mentality*. Pennsylvania: Pennsylvania State University Press, University Park.

Ehn, S. (1992), "Hur används bostaden: Kulturella aspekter på boendet i Rinkeby" ("How is a house used: Cultural aspects of living in Rinkeby"), in Ehn, S. (ed.) *Så här bor vi: om invandrares liv och boende (Such is Our Way to Live: About Immigrants' Life and Housing)*. Stockholm: Byggforskningsrådet.

Foucault, M. (1993), *Diskursens ordning (The Order of Discourse)*. Stockholm: Brutus Östlings Bokförlag.

Goldfield, D. R. (1979), "Suburban development in Stockholm and the United States: A comparison of form and function" in Hammarström, I., Helmfrid, S. and Reuterswärd P. (eds), *Growth and Transformation of the Modern City*. Stockholm: Swedish Council for Building Research.

Gustafsson, G. (ed.) (1993), *Demokrati i förändring (Democracy in Transition)*. Stockholm: Publica.

Häll, L. (1994), *Föreningslivet i Sverige – en statistic belysning (Activities of Interest Organisations in Sweden – A Statistical Analysis)*. Stockholm: Statistics Sweden.

Hanström, M. B. and Johansson, R. (1994), *Rinkeby – förort i förvanling (Rinkeby – A suburb in Transition)*. Stockholm: Byggforskningsrådet.

Hernes, H. (1987), *Welfare State and Women Power: Essays in State Feminism*. Oslo: Universitetsförlaget.

Holm, T. and Khakee, A. (1994), "Invandrarna och den kommunala planeringen" ("Immigrants and municipal planning"), *Plan*, 48(4): pp. 191–195.

Khakee, A. and Johansson, M. (1999), "Not on our doorstep: Immigrants and 'blackheads' in Sweden's urban development", in Khakee, A., Somma, P. and Thomas, H. (eds), *Urban Renewal, Ethnicity and Social Exclusion in Europe*. Aldershot: Ashgate.

Lewin, L. (1994), "The rise and decline of corporatism: The case of Sweden", *European Journal of Political Research*. 26 pp. 59–79.

Lindén, A. L. and Linberg, G. (1991), "Immigrant housing patterns in Sweden" in Huttman, E. D. (ed.) *Urban Housing Segregation of Minorities in Western Europé and the United States*. London: Duke University Press.

Micheletti, M. (1998), "En demokratisk revision av organisations-Sverige" ("A democratic evaluation of Organisation-Sweden"), *Synliga och osynliga vinster, kooperativ årsbok (Visible and Hidden Gains: Yearbook of Cooperative Society)*. Stockholm: Föreningen Kooperativa Studier.

Molina, I. (1997), *Stadens rasifiering. Etnisk boendesegregation i folkhemmet (Racialization of the City. Ethnic Residential Segregation in the Swedish Folkhem)*. Geographical Regional Studies No. 32, Uppsala: Uppsala University.

Murdie, R. A. and Borgegård, L. E. (1998), "Immigration, spatial segregation and housing segmentation of immigrants in metropolitan Stockholm 1960–95", *Urban Studies. 35*(10): pp. 1869–1888.

OECD (1996) *Housing, Social Integration and Liveable Environment in Cities*. Paris: OECD.

Öresjö, E. (1996), *Att vända utvecklingen (To Turn Back Development)*. Stockholm: SABO.

Power, A. and Tunstall, R. (1995), *Swimming Against the Tide: Progress or Polarization in 20 Unpopular Estates*. London: Joseph Rowntree Foundation.

Putman, R. D. (1993), *Making Democracy Work. Civic Traditions in Modern Italy*. Princeton: Princeton University Press.

Rinkeby Stadsdelsförvaltning (1999a), *Stadsdelsutveckling i Rinkeby 1974–1998 (Development in the District of Rinkeby)*. Rinkeby: District Administration.

Rinkeby Stadsdelsförvaltning (1999b), *Budget & Verksamhetsplan 1999 (Budget and Plan for Activities)*. Rinkeby: District Administration.

Robinson, V. (1987), "Spatial variability in attitudes towards race in the UK", in Jackson, P. (ed.) *Race and Racism: Essays in Social Geography*. London: Allen and Unwin.

Rodrigo Blomqvist, P. (1997), *Vem representerar invandrare? En studie av kommunalpolitiker i Göteborg (Who Represents Immigrants? A Study of Municipal Politicians in Göteborg)*. Göteborg: Department of Political Science, Göteborg University.

Rothstein, B. and Bergström, J. (1999), *Korporatismens fall och den svenska modellens kris (Corporatism Fall and the Crisis in the Swedish Model)*. Stockholm: SNS Förlag.

Rothstein, B. et. al. (1995), *Demokrati som dialog (Democracy as Dialogue)*. Stockholm: Democracy Council's Report, SNS Förlag.

Sangregorio, I. L. (1989), *Rinkeby: Bygg inte bort livet (Rinkeby: Do not Chase Away Life)*. Karlskrona: Boverket.

Statitstics Sweden (1996), *Befolkning, sysselsättning och pendling (Population, Employment and Commuting)*. Örebro: Statitstics Sweden.

Ungdom mot rasism (1995), *Vit makt? En studie av invamdrares och deras barns representation inom politik, förvaltning, näringsliv, organisationer och media (White Power? A Study of Immigrants and Their Children's Representation in Politics, Public Administration, Interest Organisations and Media)*. Stockholm: Ministry of Public Administration.

Westerståhl, J. and Johansson, F. (1981), *Medborgarna och kommunen. Studier av medborgerlig aktivitet i representativ folkstyrelse (Citizens and Local Government. Studies of Civic Activities in Representative Democracy)*. Stockholm: Communal Democracy Group, Ministry of Public Administration.

Westin, C. and Dingu-Kyklund, E. (1997), *Reducing Immigration. Reviewing Integration. An Account of the Current Facts, Figures, and Legislation Concerning Multiculturalism in Sweden. The Swedish RIMET Report 1995*, Stockholm: CEIFO.

Wirtén, P. (1998), *Etnisk boendesegregering – ett reportage (Ethnic Housing Segregation: A Report)*. Stockholm: Boinstitutet.

Acknowledgements

This chapter has been prepared as part of the Research Project entitled "Political exclusion in a city: The Case of Örebro", financed by the Swedish Council for Working Life and Social Research.

Table 7.1   Occupancy of different types of housing (percent)

Type	Swedes	Immigrants	
		Migration before 1975	Migration after 1975
Owner-occupied houses	55	37	14
Tenant-owned and co-operative flats	11	13	9
Rental apartments	29	42	63
Other (subletting, employer provided, etc.)	5	8	14

Source: Andersson-Molina, 1996.

Table 7.2   Segregation index of certain ethnic minorities by year of migration

Country of origin	Immigration prior to 1985	Immigration after 1985
Turkey	0.69	0.77
Lebanon	0.68	0.73
Ethiopia	0.62	0.69
Iraq	0.65	0.68
Chile	0.53	0.64
Iran	0.52	0.58
Poland	0.37	0.50
Finland	0.30	0.40
All Foreign-Born	0.25	0.47

Source: Andersson, 1998.

Table 7.3   Membership and level of activity in interest organizations (percentage of adult population)

	Swedes	Swedes	Immigrants	Immigrants
	Active	Passive	Active	Passive
Trade unions	15.1	68.4	13.0	65.6
Ideological organizations	1.5	2.5	1.0	1.5
Identity organizations	2.9	13.1	2.3	9.1
Leisure organizations	9.2	7.5	2.2	3.7

Source: Häll, 1994.

Table 7.4   Proportion of immigrants in the total population and in the tenants' organization (according to the area of origin)

	Entire Immigrant Population	Nordic	Rest of Europe	Rest of World
Population	18%	7%	6%	5%
Tenants' organization	16%	8%	8%	0%

Source: Ungdom mot rasism, 1995.

Table 7.5   Rate of employment (percentage of the population in working age)

	Immigrants	Swedes
Örebro	37.6	70.1
Brickebacken	29.7	44.7

Source: Statistics Sweden, 1997.

Table 7.6a Views on representation and influence and nomination procedures by passive immigrant members

Representation	Influence and Nomination Procedures
Under representation is considered a problem because there are no mediators between ethnic minorities and the Swedes.	Strong feeling of resignation since proposals from ethnic minorities are never really taken into account.
Imbalance in the representation enhances conflicting views and creates "we – others" feelings.	Minorities' proposals often dismissed as being too unrealistic or unimportant.
Tendency to lump all ethnic groups in a single immigrant population enforces "we – others" feelings.	Tenants' associations are bad at providing information on their activities and even more so about possibilities to take own initiatives.
Tendency to put all the blame on ethnic minorities when things go wrong in the neighborhood.	Strong tendency to arrive at "general" solutions and neglect differences between various ethnic groups.
Uncertain if increased representation would really matter since the Swedes would not listen to the views of the minorities anyway.	Avoidance of harsh discussions leads to the neglect of issues of interest for the ethnic minorities.
Lack of good knowledge of Swedish language is a major hindrance in participation.	
High unemployment rate among immigrants leads to despair and subsequent lack of interest in civic activities.	

Table 7.6b   Views on representation and influence and nomination procedures by active immigrant members

Representation	Influence and Nomination Procedures
Appointment to the executive committee depends on personal acquaintance with someone in the committee.	While representatives of ethnic minorities are free to come with proposals at the committee meetings, it is often obvious that the Swedes pay less attention to immigrants than when fellow Swedes come with proposals.
There was an expressed desire to have some women with ethnic background in the committee.	Lack of proficiency in the Swedish language is an obstacle for ethnic representatives in their work in the executive committee.
None of the elected representatives feel that they "represent" ethnic minorities, however, ethnic minorities see them as their representatives.	Cultural differences are another obstacle.
Ethnic representation is important because it enables one to understand the Swedish rules and regulations.	
There is a need to apply a quota system so that at least two persons from ethnic minorities are represented in the executive committee.	
The principles of equality in representation between men and women should also be applied in the case of minorities.	

Table 7.6c   Views on representation and influence and nomination procedures by active Swedish members

Representation	Influence and Nomination Procedures
The committee's composition is ok even though there are no representatives of ethnic minorities in the district committee. At the same time it would make it easier to solve conflicts if minorities were represented.	The election committee has the power to determine who shall be elected in the executive committee.
Cultural difference (e.g., differences in bedtime among Swedes and non-Swedes) and proficiency in Swedish act as obstacles.	Interest and commitment are two important criteria, moreover a representative composition is also important.
Lack of feelings for voluntary associations and attitudes towards the work and activities of these associations are important reasons why ethnic minorities are not represented.	Normally those in the committee are requested to continue. However, if new members are nominated then personal acquaintance is important. It is necessary that the committee knows who the new members are.
Swedes live after their almanacs and have long-term planning whereas ethnic minorities tended to be more spontaneous.	Social networks are crucial in getting commissions of trust. In a neighborhood, people know each other. They often suggest a person they know as the right person to be an executive official.
All who want to get elected have the same opportunity. There is no need to make use of a quota system.	

Chapter 8

Ethnic Minority Communities and Urban Renewal in Nottingham[1]

RICHARD SILBURN

Introduction

For some years now, "partnership" has been a buzzword in public policy planning both within the UK and in the European Union. An early expression of the case for partnership between planners, policy-makers and the public can be traced back to the Skeffington Report "People and Planning". Published in the UK in 1969, this report made the case for greater public participation in the processes of urban planning. Closer working relationships between different agencies and departments was attempted in the UK in the 1970s with the Inner City Partnerships. They were "a central plank of government urban strategy… however, those partnerships were designed primarily to co-ordinate public sector responses to urban decline and to integrate the efforts of national and local government departments. The community sector was involved only at the margin, and the private sector was a relatively minor player" (Geddes, 1997, p. 6).

Since then, ideas about partnerships have gradually become broader and more inclusive. "In the sphere of urban regeneration, the focus has shifted from a narrow preoccupation with physical regeneration to a wider concern with the economic and social regeneration of communities. Social as well as physical investment has become an important element in partnership programs, and this has brought with it the involvement of 'community' interests alongside the public and private sectors" (Geddes, 1997, p. 7).

Inter-agency partnerships, involving several departments of local (and now regional) government, working with other public sector agencies, such as the police and health authorities, may encourage a "more holistic and strategic approach to tackling problems at the local level. Partnership working can enhance both private and public sector leverage. Other benefits include better co-ordination, the reduction in duplication of effort (and thus reduced management and administration costs) and efficiency through the synergy of partners working together with a different blend of skills" (Brennan, et al 1998, p. 1).

Both the City Challenge and the Single Regeneration Budget (SRB) programs of the 1990s were centered around local partnerships, which characteristically included local government, other public agencies, local business interests, and voluntary sector organizations including residents groups. But current thinking on regeneration issues stresses the importance of an even more inclusive partnership model. This would involve local people (including residents, business people, and those who work in a neighborhood) in new partnerships with one another and relevant public authorities, to help develop and realize strategies for social and economic regeneration.

Since the election of 1997, the theme of partnership has been a central part of the Labour government's thinking. The Labour Party manifesto talked of moving from a "contract culture to a partnership culture". At the heart of this approach is the view that much public policy fails to achieve its own goals because of the extreme departmentalism of public administration at all levels. This inhibits co-operation between one department and another, and encourages what is criticized as a "silo-mentality". The plea for "joined-up thinking", leading to "joined-up action" is a recurrent theme in New Labour discourse, and has strongly influenced proposals for the reform of Local government. It is especially marked in the evolving strategy for urban regeneration and neighborhood renewal. A framework document for the National Strategy for Neighborhood Renewal was issued in April 2000 (SEU 2000). This strategy draws on the work of no fewer than 18 Policy Action Teams (PAT), one of which, PAT 17 was specifically concerned with "Joining it up Locally" (DETR 2000). What the Strategy proposes is "clear and inclusive mechanisms for joint working at national, regional, local authority and neighborhood levels, and a commitment to involving those outside government in neighborhood renewal" (SEU 2000, p. 74). At the local authority level, this would be through a new Local Strategic Partnership (LSP) "that would help services to work with each other, with communities and with the private and voluntary sectors" (SEU 2000, p. 78). The LSP would also have an important part to play in the new Community Planning process. The model is one of a vertical hierarchy of partnerships, at national level between departments of central government and other national agencies, at regional level involving the Government Offices for the regions, the Regional Development Agencies, major employers and business interests, local authorities etc, and at local authority level through the broadly-based membership of the LSP. At each level there will be a corresponding horizontal network of partner organizations, agencies and groups.

But what about the level below the local authority? Incorporated into the work of the LSP will be many local partnerships that operate in just one neighborhood, district or on one estate. For PAT 17, this local engagement is crucial. *"Empowerment is essential*: unless the residents of deprived communities are partners in joint working, nothing will change" (SEU 2000b).

This chapter is essentially concerned with these "grass-root" partnerships, which provide the essential bridge between policy-planning and bureaucratic action, and the world of ordinary people living their lives. It may be that this is the most brittle link in the partnership chain. Active participation at all levels from the local authority upwards will be mainly by individuals who will be officers, employees or representatives of a participating organization or agency. This will be true for some of those engaging at the neighborhood level too, but it is only at this level that a major contribution is asked of people who represent no-one but themselves, who are there in their own right as concerned residents, voters and citizens.

It is at this local level that many ordinary people first become engaged in community-life, sometimes as public-spirited individuals acting alone and sometimes through their involvement with a tenant or residents group, or some other relatively informal community group. Where there are already robust social networks linking people with one another, and where it is normal for people to join together in community groups and voluntary organizations to address common problems and interests, then neighborhood partnerships may be easier to establish and sustain.

Successful local partnerships, bringing together local residents and members of local groupings, will play an important part in developing social capital, both in terms of individual capacity-building and by helping to create community trust and solidarity. But they may also be prey to local animosities, personality conflicts, and may become a locus for other local tensions and social divisions. To succeed, partnership requires a degree of trust between participants, and trust must extend beyond the neighborhood partnership to the LSP and so on upwards through the partnership hierarchy. Building that trust is the first and most persistent challenge to be faced by a culture of partnership.

Problems of this kind confront all neighborhood partnerships, even the most long-established and homogeneous, but may be especially troublesome in neighborhoods characterized by social diversity, where language and cultural differences are to be found, and in neighborhoods with severe problems of deprivation and social exclusion.

Hyson Green and Forest Fields are two adjacent and severely and multiply disadvantaged inner-city neighborhoods in Nottingham. Both neighborhoods have a very diverse ethnic and cultural mix, which can be at the same time a source of tension and social division and of considerable strength and energy. This chapter explores the extent to which local residents are involved in social networks and community groups of all kinds, and whether these include people from the local ethnic groups. What are the barriers to more effective partnership working, and in particular what is the scope for and what are the barriers to greater and more effective multi-ethnic involvement in local social and economic planning processes.

The Context

Greater Nottingham is a conurbation of about 500,000 people, 130 miles north of London. At the heart of the conurbation is the City of Nottingham, the county town of Nottinghamshire, and accepted as the regional capital of the East Midlands. The City of Nottingham is a major commercial, administrative and cultural center. The City has a population of 260,000 people, about half that of the conurbation, and the City's boundaries adjoin those of four local government District Councils, each jealous of its own powers and suspicious of its Neighbors. This makes coherent economic and social planning very difficult. For the City this is particularly the case, as the tightly drawn boundaries mean that the City Council administers a large, dynamic and extremely successful commercial, cultural and administrative business center, surrounded by relatively deprived inner-city residential districts, and an outer ring of large, municipal estates of social and Right to Buy housing. The more prosperous suburbs lie in the main beyond the City's boundaries. This is an unsatisfactory social balance, creating severe problems for local planning, revenue raising and service-delivery.

By English standards, Nottingham is a culturally and ethnically mixed population. The great majority of Nottinghamshire's ethnic populations live in or near the City of Nottingham, and account for nearly 11 percent of Nottingham's population. The major ethnic groups in Nottingham are African-Caribbean (about 12000 in 1991), Pakistani (about 7000) and Indian (about 5000). There are a further 2000 other Asians, including Vietnamese, Chinese etc. In addition, there are people from other European backgrounds, including sizeable groups of people from Polish, Ukrainian, and Italian origin (who first arrived during the 2nd World War), and there has always been a pattern of internal migration, particularly from Scotland, Wales and Ireland.

Looking more closely at the situation in Nottingham itself, it is immediately apparent that the ethnic groups are concentrated in certain neighborhoods. The 10 wards in the entire County with the largest ethnic populations are all within the City. The four wards with the highest proportion of ethnic populations, in the range 22–35 percent, are all in the inner city. The African-Caribbean population is heavily concentrated in two inner-city wards (where it accounts for between 11–14 percent of the population), and more thinly distributed across a number of others. The Indian population is more widely distributed, nowhere reaching more than five percent of the local population, and the areas of relatively high residence include two of the most prosperous wards in the City, reflecting the numbers of prosperous business-people and professionals in that group. The Pakistani population are much more tightly concentrated in and around two inner-city neighborhoods, one to the north and the other to the east of the city-center, where they account for up to 15 percent of the total population. The neighborhoods of Hyson Green and Forest Fields are located in Radford and Forest wards, the wards with the second and third largest concentration of ethnic populations in either City or County.

The Neighborhoods: the Place

Hyson Green and Forest Fields are adjacent inner-city neighborhoods about one mile to the north of Nottingham city-center. Forest Fields is predominantly a neighborhood of late Victorian and Edwardian terraced housing, with a few streets of later date and more substantial construction. Hyson Green has a mixed housing pattern, as many of the older terraces, at one time typical of the neighborhood, have been redeveloped.

The housing tenure-pattern in both neighborhoods is a mixture of owner-occupied houses, and Council, Housing Association, and privately rented properties. Housing conditions are variable, and the pattern of mixed tenure has resulted in uneven repair and maintenance. This is particularly true of the private sector. Some owner-occupiers are unable or unwilling to afford the costs of normal house-maintenance. The worst housing conditions are to be found in the privately rented sector. Some landlords neglect all but the most rudimentary maintenance of the properties they own.

Housing and population densities are high; Forest ward has the highest proportion of overcrowded households in the City, and Radford ward the second highest. This can give rise to stress within and between families. More generally, overcrowding exacerbates environmental problems; problems of street litter, rubbish disposal and dog fouling are a source of constant and corrosive complaint among residents.

All the standard indicators show that Hyson Green and Forest Fields are neighborhoods marked by longstanding and persistent multiple disadvantage. In 1994 the County Council published its report "Social Need in Nottinghamshire" in which both areas were described as in "extreme social need" (Nottinghamshire County Council 1994). Both have endured persistently high rates of unemployment, contributing to poverty and financial insecurity. For years Radford ward has had either the highest or the second highest ward rate in the city, and Forest ward the second or third highest. Unemployment in Radford ward is consistently two and a half times greater than the rate across the city, and between three and four times the county rate; Forest ward has an unemployment rate that is around twice the city average, and three times the county figure.

Figures produced by the Nottingham Health Authority show that Radford ward has the highest (i.e. the worst) Jarman score for any ward in the City or the Health District, and only one ward has a higher Townsend score. Forest ward is not much better, being 96th out of 104 health district wards for the Jarman score and 97th for the Townsend score (these are well known indices of deprivation).

Both neighborhoods also suffer from a severe negative image. During the 1980s they (and especially Hyson Green) acquired an exaggerated but nonetheless unsavory reputation for violence, crime, prostitution and drug abuse. To some degree this image persists. But there is a widespread local conviction that it is not only unjustified, but that it has real and damaging effects on the lives and interests of local people. It is alleged that employers are reluctant to offer jobs, or even interviews, to applicants from an NG7 address; that it is harder to obtain credit; insurance costs are higher; that local morale is damaged, and people in other parts of the City may be put off from moving into Hyson Green and Forest Fields, except as an unwelcome last resort.

Less well-documented are the many positive features and sources of strength and potential strength that these neighborhoods have. There is a wide range of local community amenities and facilities, close at hand and accessible. There are several community and neighborhood centers with organized activities and facilities, and with clubrooms, which can be hired for private parties or celebrations. Within the neighborhoods there are several primary and junior schools serving the local population, and one secondary school. There is a public library with a strong local identity and associations; this is heavily used and highly valued. The bus-services to and from the City are excellent, and the proximity of the neighborhoods to the City's shopping and leisure services is widely appreciated.

In recent years the range of shops has declined severely, and those that remain are commercially vulnerable as supermarkets and one-stop shops have changed shopping habits and transformed the retail sector. Nonetheless local shops and markets in Hyson Green still compare well with what is available on most housing estates. There is a large Asda supermarket, which is used regularly by most people from both neighborhoods. But there are also a significant number of small shops, including butchers, greengrocers and newsagents, as well as the many charity-shops, all providing some degree of consumer-choice. In the last few years there has been a noticeable increase in Asian-owned and run shops; some of these supply the particular needs of the Asian communities for clothing or jewelry for example, or for Halal meat. But others serve a much wider cross-section of the community, with fresh fruit and vegetables and household goods. Similarly the Chinese and Indian take-away restaurants, and the Indian sweet-shops have joined the traditional fish and chip shop as amenities enjoyed by most sections of the community, and probably by all the younger people in the neighborhood. There is no doubt that the survival of a diverse retail sector in these neighborhoods owes a great deal to Asian entrepreneurs. There is some evidence to suggest that successful niche marketing is gradually adding to the range of local shopping opportunities, and is attracting shoppers from other parts of the City. Hyson Green is well served by pubs, some of which have, mercifully, not yet been modernized.

The spiritual needs of the neighborhoods are met through the local Christian churches and chapels, the three mosques in Forest Fields and the Sikh and Hindu temples within the neighborhood or nearby.

There is street-life, and large numbers of children still use the street as a playground. This close mix of homes, schools, shops, social and other commercial facilities give Hyson Green/Forest Fields its distinctive character. Despite widespread severe poverty and disadvantage, these are not seriously demoralized neighborhoods. Individuals live under varying degrees of stress and pressure, but this is not reflected in a widespread rejection of the neighborhood, or withdrawal from participation in local activities and events. There is considerable local pride and loyalty. Many residents positively like their neighborhood, and see many advantages and benefits from living there that greatly outweigh the disadvantages. They remain residents as a matter of choice. In a recent community profile survey of 311 residents of Hyson Green, three-quarters of the respondents said they would like to stay in the area. They liked the area, felt at home in it, liked the people, appreciated local facilities and the closeness to the city center.

Many people feel that it is a vibrant community, precisely because it is not a single homogeneous community, but an extremely diverse pattern of distinct but overlapping communities each with their own energy and interests. Nonetheless, the high levels of persistent unemployment and widespread poverty mean that many residents of whatever ethnic or cultural background are socially excluded and have been so for many years. But for some of the ethnic minorities there are additional layers of exclusion, some of it self-exclusion, based on differences of color, of culture and beliefs, of social custom and convention.

The Research Neighborhoods: the People

The population of Hyson Green and Forest Fields is a more complex social mix than first impressions might suggest. There is a mixture of residents of longstanding and newcomers. The greatest turnover of population is in the privately rented sector, where short-term tenancies are the norm. In recent years there has been a substantial increase in the population of students staying for no more than one or two years. There is considerable "churning", as individuals and families come and go. A high level of population turnover has a destabilizing effect; people have less opportunity to get to know one another, social networks are weaker. The work of the local schools is disrupted by constant turnover of pupils. Greater population stability might be one indicator of local regeneration.

There is a small but significant middle-class presence particularly in the streets with larger houses to be found around the edges of the research neighborhood.

The Ethnic Mix

The population is both ethnically and culturally mixed. Hyson Green and Forest Fields have been ethnically mixed neighborhoods for many years, and there is a local acceptance that this is, and will remain the case. In this respect Hyson Green and Forest Fields are good examples of working multi-cultural communities. Everyone born and bred in Hyson Green and Forest Fields under the age of thirty have lived all their lives in an ethnically mixed environment. They will have attended ethnically mixed schools, and had the chance to mingle with people from different races and backgrounds. Widespread street-racism seems to be, in general, a thing of the past.

But it is also the case that many people in all the ethnic minority groups still feel significantly excluded from the mainstream of British life. They are convinced that there is widespread and deep-seated institutional racism that disadvantages all ethnic minorities. This is particularly problematic in the area of employment. Again and again we were told of applications for jobs foundering when an ethnically distinct surname was mentioned, of the problems of securing credit, or insurance cover, and many other such examples of discrimination. For all the disadvantaged people of Hyson Green and Forest Fields, and especially the ethnic minorities, mainstream England continues to be an unfriendly, potentially hostile environment.

The term "ethnic minority" can be misleading in two ways. First it may suggest that there are differences between one ethnic group and another that are greater than is actually the case. Second it may hint at a greater degree of internal cohesion and cultural similarity than is justified. In fact most of the ethnic minority groups are far from homogeneous.

The African-Caribbean population (two-thirds of whom were born in Britain) came originally from the West Indies but from several different islands separated in some cases by hundreds of miles of ocean. People from different islands who may never have encountered one another before coming to England will not automatically feel that they have a great deal in common, and may not feel that they share a common culture until one evolves from their shared experience of life in England. Indeed it is sometimes the case that there are inter-island rivalries. For this group, there is no one common religion, a single uniting faith which asserts authority and commands universal respect, and which contributes to a broad sense of shared identity. Unlike some other ethnic groups, African-Caribbean are English-speaking, so there is no severe language barrier to separate them from the host population, and bind them to one another.

The Indian community is another heterogeneous grouping. Those who came to England directly from India reflect some of the enormous diversity of that vast country; they speak different mother tongues, they hold different religious beliefs, they have different dietary habits. Others came to England, not from the Indian subcontinent but from East Africa, from a colonial and post-colonial environment in which they had had a major role as traders and business people.

The Pakistani community is significantly more homogeneous than either the African-Caribbean or the Indian. This is partly because they are the most recent sizeable group to have arrived in England; the first major wave of immigration from Pakistan into Nottingham did not occur until the 1960s. But more importantly, the great majority of both initial and subsequent arrivals came from one small autonomous region in Pakistan, the Mirpur district of Azad Kashmir. They brought with them, and have retained, extensive and close ties of family and clan. However, Kashmir is a divided and unsettled region with its own history and traditions, and many Mirpuri Kashmiris feel themselves to be different from people from other parts of Pakistan.

Ethnic Groups in Transition

The African-Caribbean Population

For many of the older generation of black people, the reality of life in England has not matched their original high hopes. What many felt to be the mother-country turned out to be less welcoming than hoped for, and the pervading sense of being excluded by prejudice and discrimination runs deep. One reaction to this sense of rejection has been the gradual evolution of a black awareness. For the older generation this was inspired by developments in the United States. There is now more than one generation of younger, British-born African-Caribbeans, and although the celebration of blackness remains an important cultural feature, there are many young black people who are striving to realize their identity as Black British, with a distinct contribution to make in an evolving multi-cultural Britain.

The Indian Population

The small Indian community appears to have adapted most readily to life in Britain. They had certain advantages to start with; the first generation of Indian immigrants were more likely to have come from an urban background, were better educated, and brought with them more transferable skills, including professional skills, than was the case with other groups. East African Indians were members of a merchant and business class of considerable experience, sophistication and self-confidence. They see themselves and are seen as hardworking, ambitious and entrepreneurial, as a group both self-confident and successful. This group of people occupy a different segment of the occupational and class structure than is typical of either the black or Pakistani populations.

The Indian community is better educated and wealthier than other ethnic minorities, has a stronger and more mainstream business and professional exposure. There is a sizeable Indian middle-class, which is ambitious and achievement oriented, and is both upwardly and geographically mobile, with many moving to live in more prosperous areas both within the City and beyond as their circumstances permit.

There is now a second generation of young British Indians. Those from relatively comfortable and supportive backgrounds, with reasonable (and sometimes exceptional) educational achievement, seem to feel comfortable living as it were in two cultures at once. They do not feel as alienated and excluded from mainstream British life as do many young African-Caribbeans, nor is the parent culture felt to be as restrictive as it is for some of the young British Pakistanis.

The Pakistani Population

The population of Pakistani origin is the largest ethnic group in the research area, and the most cohesive. In contrast to the African-Caribbean immigrant population, who were English-speaking, frequently with strong and shared Christian beliefs, and who felt (rightly or wrongly) culturally closer to the mother-country, the first generation of Pakistani residents had language barriers which made communication with other people extremely difficult, and deeply-felt religious beliefs, patterns of worship and dietary habits which were unfamiliar to the host population. The first immigrants often lacked formal education or transferable skills, and some were illiterate. These factors conspired to set them apart from other groups within the community, and made it difficult for them to be able to play a full part in the life of the wider community. For these reasons, social integration within the wider population has been relatively slow.

All this is changing, and a process of gradual adaptation and transition is now well under way. While language difficulties are still a barrier for many of the older members of the Pakistani community, there is now a second generation that is more likely to be bilingual. For the second-generation links with Pakistan remain strong and important, but most recognize that they are unlikely to return there to live, and they expect to remain living and working in the UK.

The networks of family and faith are a strong defining characteristic, around which have developed extensive patterns of self-help and mutual aid. In a district marked by rapid turnover of population, the Pakistani community is now an important stabilizing element, a large, linked population who remain in the neighborhood because they choose to, because it meets their most important needs and concerns. There are a number of strikingly successful individuals who could easily afford to move to other residential areas where they could buy newer, larger and better-equipped properties, but who choose to remain as residents in or near Forest Fields, because of the importance they and their immediate family attach to maintaining the extended family connections and networks in the neighborhood.

Considering the obstacles that had to be overcome, what that second generation has accomplished is remarkable, and it is clear that there is now an emerging and more self-confident Pakistani entrepreneurial and business class, often self-employed (whether by necessity or preference), often in the service sector.

There is now a rising third generation, the grandchildren of the first immigrants. These young people are predominantly British-born and raised, and are bi-lingual. They have attended English schools throughout their school-lives, and are able to mix more easily with children from other ethnic backgrounds. The links with Pakistan are still important, but they are weaker. Many have never visited Pakistan, even for a holiday, and there is no serious question of them ever returning to

Pakistan to live. They (and their parents) recognize that their upbringing and experience are such that they would not be able to make the considerable adjustments to daily life in rural Pakistan that would be necessary.

For better or worse, these young people see themselves as British Asians, and they too must strive to realize a distinctive British identity, that enables them to remain as valued members of the (changing) Pakistani community while also being accepted as members of a multi-cultural, mainstream society. This is not an easy phase and for many of this third generation there is something of a crisis of cultural identity. Family ties remain strong, and the cultural values of their parents and grandparents are a strong influence; but at school and with friends from different backgrounds, they encounter different cultural assumptions and patterns of expected behavior. They are able to move more easily between one cultural setting to another, to shift between the Asian and the Western, and perceive the differences between cultures less as incompatibilities and contrasts and more as choices. For some of the older generation, there is the abiding fear that traditional cultural values and beliefs are slowly becoming diluted and weakened. This sometimes is a cause of tension and difficulties within families and between the generations, particularly where there is a fear that religious values are threatened or undermined. For the older generations, family, faith and neighborhood are a source of strength and security, but some of the younger generation may have more ambivalent feelings, and what some see as security others see as a trap. The emotional security of family and clan ties may be complicated by its controlling function, and individuals may press for a greater degree of personal autonomy.

Some prominent members of the Pakistani community argued that neither nationality nor culture, are fundamental here; nationality is linked with geographical place, and becomes less identity defining in a world of constant mobility. Cultures are not fixed patterns of thought or behavior, but are constantly evolving, adapting to constantly changing circumstances. But, it was argued, what is unchanging is faith, and for these British Asians the eternal truths and values enshrined in Islam should continue to be a defining force. The greatest danger they face is to lose contact with the faith, and become rudderless in a potentially hostile world. It is certainly felt that an important part of the work of the mosques is to help the rising generation maintain their faith. But this is not made easy in an environment where ignorance, fear of the different and religious intolerance is more widespread than blatant racial discrimination.

An added uncertainty for this young generation is economic insecurity. What job and career opportunities will they have? Will they be able to break into mainstream employment, in the evolving global job market. Will the sons and daughters of the taxi-driver or take-away owner, succeed in becoming the teachers, engineers, pharmacists, social workers, computer operators and financial advisors of their generation?

These are very complex issues; but it is clear that each of the major ethnic minority groups is in the middle of a long-term process of transition, striving to realize an identity that retains the most cherished aspects of their distinct cultures, while accepting that they are now an integral part of British society. At the same time the white population is itself coming to terms with the fact that they too live in a multi-ethnic society. The transition is well underway, but far from complete.

Social Networks, Community Groups and Voluntary Organizations

Social Networks

The phrase "social networks" refers to the links that exist between a resident in a neighborhood and other people outside that person's own household. These may be ties of family and kinship, of friendship, or between Neighbors and near-Neighbors. Informal networks are a major focus of social contacts outside the home. They act as a grapevine for information and gossip, and for finding out about what is going on in and around the neighborhood. They have practical day-to-day significance as support and mutual aid systems. They help people to find shared solutions to common problems. Those with rich informal networks are more likely to feel positively about their neighborhood, and to have a greater sense of belonging. These networks, is an aspect of social life that many people take for granted hardly worthy of comment. But patterns of family life (and family problems), of sharing between Neighbors, friends and other spontaneous social contacts are a bedrock of local community.

Family ties remain an important part of most people's social relationships. But family ties are not necessarily local ones. Relatives may live in other districts of the city, and even further afield. In these circumstances face-to- face contact is less likely to be a daily occurrence, and in many cases the telephone is the easiest way for family members to maintain contact. While this physical separation may make it more difficult for day-to-day casual support to be relied on, family would still be thought of as a first source of help in a crisis or emergency.

Amongst the Asian community, close family and clan ties within the immediate neighborhood are both extensive and complex. Research into Pakistani communities in other British cities has shown that "settlement patterns... in different towns, or on a smaller scale within towns, on the basis of their home origins came about as a result of the kin-friend chain migration...". There is a sharp focus on family and clan ties; "visits to each other's houses are quite frequent... regular meeting places include Pakistani shops, mosques, supplementary schools... the situation as a whole leads to multiplex relations and a close-knit community... among many it is not felt necessary to deal with the indigenous people due to the availability of ethnic facilities" (Anwar 1995, p. 238). This is certainly the case in Forest Fields, where the great majority of the Pakistani population originated from one district in Pakistan-administered Kashmir.

Beyond the family, there are contacts with Neighbors and near-Neighbors. These may develop into friendships, with more regular, more frequent and more social and recreational contact. The informal social web is more developed where the resident population is rooted, familiar with the neighborhood and its inhabitants, where there is a more settled pattern of life. It is weaker on the streets with the greatest turnover and churning of population. Networking is easier where individuals recognize the features they have in common with others, and more difficult where differences between individuals and families is more marked, or where the anti-social behavior of a few people throws everyone else into a more defensive and cautious frame of mind.

The essential precondition for the formation and continuation of these kinds of community bonds is trust. Trust is usually greater between people who have known each other for some time.

Many people commented that both Hyson Green and Forest Fields are not single communities but a complex of communities within communities, and this is reflected in both the informal social networks and more organized community groups. The most widespread and rapid network responses occur when some widely used amenity or service is threatened in some way. Single issue, defensive campaigns of this kind can trigger a large and spontaneous (but usually short-lived) response.

Individuals vary in the extent of their involvement in local networks. Only a relatively small number of individuals move beyond the informal social web, to take part in the work of larger and more organized community groups. This is not because of apathy but because people have many different and conflicting claims on their time and energy, and active community involvement requires both time and energy, sustained over a long period. There may also be a lack of self-confidence, and uncertainty about the way groups conduct themselves. Committees, meetings, agendas and minutes, even the formalities of speaking to a group of relative strangers, make many people uneasy.

Community Groups

Hyson Green and Forest Fields both have a long history of voluntary community group activity. There are groups of all kinds. Some are campaigning groups, which reflect and articulate local needs and demands; in some cases the campaign focuses on a single issue, in which case it may attract support from those with an interest in that issue from all over the neighborhood and beyond. Other groups are self-help organizations providing or sharing services among their members. Some are both campaigning organizations and service providers. Very often it is the informal social networks that act as gateways into wider associations, and are conduits for feeding back into the community news and ideas generated in these more organized settings.

Street-based groups that form to meet immediate local needs, like some residents groups, are often informal in their methods of operation. They have no significant funding (although the City Council offers a modest start-up grant to Residents Groups), they employ no staff, they occupy no premises. The most active members may in effect form a management committee, although they may not use that term, and may think of themselves as a much more spontaneous collective. Meetings may well take place in one or another member's house, without a fixed agenda or clear record of the proceedings. The focus of activity tends to be immediate, addressing specific problems that afflict the group's catchment area.

Community groups vary considerably in size. But, large or small, they usually depend upon the energy and commitment of a minority of active members, who do most of the day-to-day organizational work, and provide the driving-force and leadership that sustains the group. Behind these activists are a larger and more passive membership, who may be prepared to take a more active part in the work of the group from time to time, but who do not involve themselves on a day-to-day basis, and whose interest may wither if difficulties arise. Women take a very active part in community groups, indeed they are often the driving-force, and most active members. This is particularly true of many of the residents groups, and those groups that are concerned with such issues as street-safety and security, and provision for child-care and young people.

Over-dependence on a very small number of active participants is one reason why some voluntary groups are short-lived and liable to sudden collapse. Nonetheless, engagement with local community groups gives many people a sense of constructive purpose and fosters individual and communal enrichment. Community activists are often involved in the work of several organizations at once. Such people come to form what amounts to a network of their own, and their experiences inform their attitude to one another and to formal service providers (especially the local authority). But the complaint is sometimes made of seeing the "same old faces".

Many community activists have been engaged in activities of this kind for a long time, often for many years. They have a considerable range and depth of experience. But those with a long history of involvement in local groups carry with them the lessons of their experience. This sometimes leads to deeply-held suspicions of the motives of other people or groups, and there can be a sharp antagonism between members of community groups and the local authorities charged with local social policy and service delivery.

There is some overlap and duplication of groups, and although the advantages to be gained from co-operation between groups with broadly similar objectives may seem obvious, there are a number of obstacles that inhibit such co-operation. These may include ideological or personality conflicts, inter-group rivalries, and disagreements about priorities and tactics. Competition between groups for scarce resources may create mistrust between one group and another.

Most community groups have immediate goals and targets. These may encourage short-term tactical thinking and action. Involvement in longer-term and strategic local planning and policy-making poses different challenges and requires different skills. But there are examples of well-established local groups, which take a leading interest in longer-term issues.

Community Groups and the Ethnic Minorities

Ethnic minority involvement with local community groups is a rather more complicated story. There are a number of very active organizations run by and for the different ethnic minorities. These include community centers, artistic and cultural organizations, recreational and sporting groups, and religious groups based around a church, temple or mosque. But it is often the case that these organizations are city-wide, or even county-wide, in their focus, and if they happen to be based in Hyson Green and Forest Fields (as is the case with the Indian Community Center, and African-Caribbean Families and Friends (ACFF)) this is coincidental. Active membership of a citywide organization has the effect of taking energy and drive out of the immediate neighborhood. In this case activists will be working not with Neighbors or near-Neighbors, but with other members of their own ethnic group, drawn from all over the city. While taking some energy out of the immediate neighborhood, these wider contacts also act as an important bridge between neighborhoods and the wider urban environment of which they are a part.

This of course does not prevent individuals from joining area-based groups, and some certainly do. They join residents groups, and other organizations, especially those organized around and through the community centers. But it is probably true that they are not as involved in local groups as their local numbers would suggest. There are different reasons for this.

The African-Caribbean population is divided about the extent to which its members should become involved in local group activity outside of separate black groups. Some argue that open membership groups cannot reflect the real needs and interests of the African-Caribbean community, will probably be dominated by and run on behalf of white people, and therefore, wittingly or otherwise, oppressive. Involvement in such groups is seen as an unacceptable compromise. Others put the contrary case that it is only by becoming involved in broader, mainstream groups that the particular needs of the African-Caribbean population can be properly registered. Moreover, special needs apart, many of the problems in Hyson Green and Forest Fields affect everyone whatever their ethnicity. In these cases, race should not be allowed artificially to divide those who have shared concerns. This debate is ongoing, and has yet to be resolved.

The networks of family and clan within the Pakistani community merge seamlessly with wider networks based around the mosque. The spirit of self-help and mutual aid continues to be very strong, and in the main, the Pakistani community continues to look to its own resources to meet its own needs. There are fewer formally organized groups, and more influence is bestowed on well-respected community and clan leaders. It is often these community leaders (invariably men) who act as go-betweens with other mainstream groups and agencies. For many of the older people, or those who have arrived more recently in England, there are still language barriers, which make participation in large English-speaking groups more difficult.

It is not yet culturally normal for Pakistani married women to take an active part in public life outside the home, and so, unlike their white counterparts, they do not play a prominent part in local residents groups or other voluntary groups in the neighborhood. They do not usually attend public meetings, and rely on male community leaders to articulate their views.

There are a number of Pakistani men and women who work in the public sector and who are more closely associated, for example, with the neighborhood and community centers, but their involvement is professional rather than voluntary, and some at least of the women will discontinue these activities on marriage.

It was suggested that for many Pakistanis there is a further cultural barrier to community group involvement, rooted in an unfamiliarity and unease with the processes that voluntary groups adopt. They are unaccustomed, it is said, to formal procedures, to agendas, and minutes of meetings, to public argument and disagreement between relative strangers. This of course is an unease that is not confined to the Asian population; for many people from all parts of the community, meetings and their associated procedures, even the challenge of speaking in a public situation are very daunting. There is some evidence to suggest that the younger Pakistanis, particularly those brought up in England, are not so intimidated or inhibited, and have mastered the basic skills of meetings and organizations more easily.

But at this stage in the process of assimilation, the Pakistani community as a whole is still predominantly inward looking, with a strong focus on self-help and mutual aid within the community. Community leaders will act as the bridge into the wider community networks, and will be trusted to defend the interests of the Pakistani populations.

So for these two major groups at least, the African-Caribbean and the Kashmiri, there are additional ideological and/or cultural obstacles to community involvement above and beyond those that are common to all socially excluded groups.

Social Exclusion and the Politics of Partnership

It is becoming increasingly recognized that for urban and neighborhood regeneration programs to be successful requires major changes to traditional ways of making and implementing policy. If there is one clear lesson that has been learned from earlier regeneration initiatives, such as Task Force, the Urban Program and City Challenge, it is that top-down, prescriptive programs have had little lasting impact on the neighborhoods they are supposed to be helping. A more inclusive participatory culture is required. So, the traditional emphasis on physical renewal, or economic development has to be complemented with as strong an emphasis on social renewal, hence the language of empowerment, social capital, capacity-building etc. At the same time there is seen to be a need for integrated and holistic local policymaking, to overcome artificial boundaries and barriers whether these arise from different agencies being unused to working with one another, or from excessive departmentalism, the "silo-mentality" within a complex organization. Hence the concern with forging multi-agency partnerships involving a wide constituency of hitherto separate interests, including voluntary sector organizations, community groups and individual tenants and residents. Strong local partnerships are needed to harness the energy, skills and resources of the key players who develop and implement local regeneration strategies. These include public sector agencies, employers, trade unions, voluntary organizations and local community groups (Geddes, 1997). This vision of partnership clearly lies at the heart of New Labor's Third Way economic and social strategy. The New Deal for Communities initiative announced in September 1998 emphasizes this more holistic approach to policymaking. The encouragement of a "mixed market" in the delivery of services, and the involvement at all stages of the widest constituency of local interests are seen as vital to successful area regeneration.

The partnership idea is an attractive one in theory but presents difficulties in practice. The list of potential partners is a long one, and embraces people and organizations with widely different interests, backgrounds and expectations. Enabling and encouraging significant numbers of people from severely disadvantaged neighborhoods into creative political partnerships has some special difficulties. In an important sense almost everyone in neighborhoods such as Hyson Green and Forest Fields, whatever their ethnic background, is socially excluded, and there is considerable overlap between the impediments to inclusion of the local white and ethnic populations. The root of the exclusion is a complex of social class-related disadvantages, including housing-stress, educational underachievement, poor levels of skill and hence job-prospects, low earnings, risks to health, family breakdown and so on. Taken together, these factors encourage a local culture which is essentially about survival, looking for immediate solutions to immediately pressing problems, mistrustful of public authorities who are often seen, not as a solution, but as part of the problem, mistrustful of one another until confidence and trust have been earned, with long memories of broken promises and hopes dashed. For the ethnic groups, there are the additional problems of real and imagined institutional racism. None of this is an ideal breeding-ground for an effective politics of partnership. There is some small recognition of this in the talk about the need to encourage local "social entrepreneurs" with the personal and organizational skills to be able to articulate local needs and wishes. On the positive side, however, there is the strong evidence of local self-help organization and community group activity. Moreover hope springs eternal, and there is a widespread wish for more effective means of political participation, tinged as it will be by considerable sensitivity and doubts, born of successive defeats, disappointments and what are often seen as betrayals by indifferent bureaucrats or self-seeking politicians.

The Dilemmas of Partnership

There are considerable disparities between the partners, in terms of legal powers and responsibilities, technical knowledge and professional expertise, familiarity with the issues to be tackled and the likely range of policy options. Some difficulties can be traced to an uneasy fit between the ethos surrounding partnership on one hand and government and administration on the other. Partnership is inclusive, discursive and collaborative. It relies heavily upon the commitment of volunteers, whose continued support depends upon healthy relationships of trust between all the partners. Government is legal and bureaucratic, hierarchical and increasingly managerial. Neighborhood partnerships operate in an environment where there will continue to be strong central policy direction and financial controls. Some partners will have links with and responsibilities to other groups and interests outside any one-neighborhood partnership. Local authorities and other public agencies such as the Police, the TEC and Health Trusts will have a leading role to play in local partnerships. Moreover in a city like Nottingham, these public agencies are likely to have such a role in a number of distinct neighborhood partnership organizations at any one time. Where bidding for funds from national or European sources is involved, these local partnerships may find themselves in competition with one another for resources, in the same way that local authorities, or the new Regional Development Agencies, will be in competition with one another when dealing with central government. The relationship between the local authority and its several partner organizations needs to be understood very clearly by all parties involved, if difficult decisions involving prioritization are not seen by unsuccessful partners as evidence of bad faith or even betrayal.

The same sort of problem might arise in the voluntary sector, where an organization might be involved in a number of competitive bidding processes, in competition with other voluntary organizations for funding support. Where a strong managerial thrust from the center is maintained, with local competition for resources, the most likely development will be a culture of "shallow partnerships", that is to say, shifting alliances between interest groups within a community, shifting in response to the experience partners have with one another, and to the steer from the center. The establishment of these shallow partnerships should be regarded as an important step towards greater local community empowerment. However dynamic and self-sustaining regeneration strategies will require deeper forms of partnership to continually refuel the commitment of the most active and committed partners.

Establishing Trust

Many partners (such as those from the business community or the local authorities) are used to working on the basis of contractual obligations or hierarchical authority. But an inclusive neighborhood partnership requires a different mode of operation, which relies heavily upon the confidence that partners have in one another and the bonds of mutual trust that evolve.

The development of mutual trust and respect between individuals and groups, who have often in the past seen one another as competitors, or even enemies, is bound to be a long process. The neighborhoods of Hyson Green and Forest Fields are within the catchment area of a Partnership Council, which has been centrally involved in helping to determine the distribution of URBAN funds, and more recently with the New Deal for Communities initiative in Nottingham. It has made some good progress, but amongst participants there remains a great deal of mistrust, suspicion and cynicism to overcome, and most, perhaps all, the partners retain some degree of suspicion about the motives of others. For example, some of the older community activists and voluntary organization personnel, as well as some residents, have many years of trade union and political activism. Many of them have a history of involvement in local political and community action, in some cases stretching back over decades. They have memories of past disappointments, of hopes dashed, and of promises broken. They remember the Community Action struggles of the early 1970s, and have watched as one top-down regeneration initiative after another has evolved; most have been in the position of applying (often unsuccessfully) for local authority funding support, or of having funding discontinued. For many, both the local authority and the business community are viewed with some degree of suspicion which can be intense. They not only mistrust the motives of the business community but also see the state, both central or local, not as some neutral force pursuing the common good, but, rather, as having powerful vested interests in its own right, not least the retention of power and authority. For such members of the community the rhetoric of "partnership" is regarded with considerable skepticism.

Mistrust of the local authority is not confined to residents and community activists. There is mistrust among many local authority employees of their own employer. They question both the will and the capacity of a large-scale, hierarchical organization, like a City Council, operating in a highly competitive managerial environment, to make a reality of any partnership which implies a genuine devolution of authority. Some councilors and local authority officers are skeptical, ambivalent and in some cases openly hostile to the very idea of local partnerships.

Sometimes the objection is one of principle, for example that the Partnership Council will usurp the authority that by right belongs to those who have sought and obtained elected public office, and who remain accountable to their local electorates. Sometimes it is a pragmatic objection, that lay partners lack the knowledge and experience to be able to make competent decisions, find it hard to reach a consensus, particularly when decisions have to be made quickly to meet deadlines, are inconsistent and, in a literal sense, irresponsible.

In brief, the politics of partnership, of collaboration and negotiation, is in many ways harder than the politics of confrontation, of winners and losers. The trust that is essential for partnership to work is in short supply, and reserves of trust will only be built up slowly. Trust grows out of experience, and positive experiences of partnership will make further even more positive experiences more likely.

The obstacles to the inclusion of ethnic minority groups in these processes are not fundamentally different from those inhibiting the inclusion of the white majority. Underprivileged and multiply deprived inner-city populations have been largely excluded from all but token participation in decision-making at any level, and are often deeply suspicious of and sometimes cynical about the rhetoric of partnership and democratic involvement. The Labour Movement has in general seen itself as engaged in a power-struggle with hostile forces, rather than as a partner in a process from which everyone can (in theory) benefit. If the politics of partnership are to become a working reality, then a profound change of mind-set is needed, and everyone who is active within a neighborhood partnership has to become part of that change. This applies equally to alienated residents, to frustrated community activists, to public officials (whose authority is threatened), and to ethnic groups trapped in the politics of the ghetto.

Conclusion

The encouragement of local partnerships lies at the heart of European and New Labour thinking, and is often described as a major "culture-shift". It promises to replace a politics of confrontation with one of negotiation, with reconciling differences by mutually agreed concessions and compromises. It marks a step away from the old top-down policy process in the direction of a more participatory style, in the direction of empowering individuals and communities. However it remains the case that partnerships will be expected to operate within a wider political environment that is essentially competitive, hierarchical and managerial.

There is as yet a lack of clarity about the precise remit and boundaries of a local partnership. Are they more than a consultative mechanism and if so how much more? What executive or decision-making powers may they have? Where do they stand in relation to other public agencies and authorities that have legally defined duties and responsibilities? These are difficult but important questions to be worked out; meanwhile, partners will come into the process with different hopes and expectations, with scope for considerable misunderstanding and disillusion.

It is widely acknowledged that partnership involving local residents in the decision-making process requires those residents to cultivate a range of skills to become more effective partners. This is no doubt often the case, but perhaps it is not sufficiently recognized that the skills deficit is itself only an indicator of much more deep-rooted, longstanding and pervasive patterns of social exclusion which have become enmeshed into patterns of local culture and expectation. These can change, but the process of erosion will not happen overnight, nor is it guaranteed. Such changes occur as a response to reassuring positive experiences, and can quickly be reversed by reinforcing negative experience. It is for this reason that the challenge of partnership may be even greater for those who have become accustomed to the exercise of power and influence in a more directive and prescriptive tradition. Local government itself is in the throes of major change. In the case of Nottingham City, the change to unitary status is still being worked through; soon the regional Development Agencies will impact, and the ideas outlined in the White Paper on Local Government anticipate major changes. Local partnership is one among many challenges to traditional working styles and methods. In the front-line will be local councilors and council officers, who have to balance the ebb and flow of local argument and suggestion with their responsibilities to the wider community they serve, while functioning within a highly competitive, deadline-driven environment. Whether they have fully understood the changes that are required of them remains to be seen. But if it should transpire that local partnerships come to be seen as little more than a manipulative device through which old power-structures continue to operate, that betrayal will take a long time to overcome.

Note

1   This chapter is based on one of four local area studies commissioned by the Joseph Rowntree Foundation as part of a long-term research program on Area Regeneration. They were all published in 1999 by York Publishing Services.

> Andersen, H. et al, *Neighborhood Images in Liverpoool 'It's all down to the people'*.
> Forrest, R. and Kearns, A., *Joined-up Places?: Social Cohesion and Neighborhood Regeneration*.
> Cattell, V. and Evans, M., *Neighborhood Images in East London: Social Capital and Social Networks on two East London Estates*.
> Silburn, R. L. et al, *Neighborhood Images in Nottingham: Social Cohesion and Neighborhood Change*.
> Wood, M. and Vamplew, C., *Neighborhood Images in Teeside: Regeneration or Decline?*

References

Anwar, M. (1995), "Social networks of Pakistanis in the UK; A Re-Evaluation", in Rogers, A. and Vertovec, S., *The Urban context: Ethnicity, Social Networks and Situational Analysis*. Oxford: Berg Publishers.

Brennan, A. et al (1998), *Evaluation of the Single Regeneration Challenge Fund Budget*. London: DETR.

DETR (2000), *Report of the Policy Action Team 17, Joining It Up Locally*. London: TSO.

Geddes, M. (1997), *Partnership against Poverty and Exclusion*. Bristol: Policy Press.

Nottinghamshire County Council (1994), *Social Need in Nottinghamshire: County Disadvantaged Area Study*. Nottingham: Department of Planning and Economic Development.

SEU (2000), *National Strategy for Neighborhood Renewal: a framework for consultation*. London: TSO.

SEU (2000b), *Policy Action Team report summaries: a compendium*. London: TSO.

Chapter 9

Minorities and Successful Stories

PAOLA SOMMA

human kind / cannot bear very much reality (T. S. Eliot)

the meaning of a word is its use (L. Wittgenstein)

It is by no means easy to identify with the approach suggested by the editors of the book, who have set out to focus on the role of minorities as catalysts of positive urban regeneration experiences for society in general. But once one has overcome the initial reaction – that is one of those pointlessly generalized invitations to positive thinking, along the lines of "don't worry, be happy", that so often underlie advertising slogans and urban marketing campaigns promoting "the city that refused to die", or "the city reborn" or "the resurrected city" and attempting to cultivate a superficial and senseless (because baseless) optimism – the ensuing reflection is neither banal nor easy.

What is required are not stories of single individuals triumphing over disadvantaged origins, obstacles and sufferings and achieving success thanks to their tenacity and enterprise. Such tales are always sought out by the mass media and presented (the more dramatic the apparent circumstances the more emphatic the tone) as evidence that hard work and a spirit of sacrifice will always win through.

The task we have been given is more intractable, not only because the reference is to groups and not individuals but also because it involves reconsidering what we mean by some of the terms and stereotypes and models that we commonly use; and most importantly of all, because it demands that we shed light on the ambivalent and contradictory relationships that can become established between the concepts of minority, local community, participation and urban regeneration.

Minorities in Urban Regeneration or Regeneration vs Urban Minorities?

The only minority groups capable of making their presence felt with an individual political strategy may be in a minority from a numerical point of view but they are nevertheless powerful, either because they are rich or because they are, as they say these days, "protected by the international community". I do not see their actions as giving rise to general or equally distributed benefits. On the contrary the destruction of national states to the advantage of so-called local communities is taking us all towards a new condition that might simultaneously be described as both pre-modern and post-modern, which embraces both feudal-type social relationships and the global market.

A similar argument can be made at an urban level. If a minority, whatever its connotations, is a group of people who suffer disadvantages compared with another, not necessarily bigger group in the same context, it is true not only that the notion of a minority is a social construction but also that the notion is not an absolute condition. There can be no minority without a majority and it is the difference or relative distance between the two groups, that may be quantified in different ways depending on the values, and therefore the indicators that prevail in any given society (we value what we gauge), that makes it possible to identify the existence of a minority and describe its characteristics.

In the second place, the disadvantage of a minority (and it is merely common sense not to confuse a minority with "the happy few" who may be small in number but who certainly do not constitute a minority) derives, at least in part, from discrimination on the part of institutions (or the disadvantage may be perpetuated or accentuated by discrimination); one of the clearest indicators of the existence of a minority is therefore a disparity in the enjoyment of rights that belong to all members of a society. Assuming unequal rights as the determining criterion for definition of a minority, it is therefore reasonable to regard non-European immigrants in Italy, in the current situation, as a minority compared with Italian citizens; and in various local circumstances they constitute a minority with respect to the other inhabitants.

This objective condition of comparative disadvantage is further aggravated by the fact that, unlike other weak groups – from the chronically sick to the handicapped and from one-parent families to the elderly – to be encountered in the increasingly crowded universe of the underprivileged, only immigrants are perceived as a threatening minority and as such arouse hostile and often violent reactions.

The so-called "immigration issue" (an abstract concept compared with the concrete problems of the immigrants themselves) is now a political question at a national level, though the prevailing tendency is still to treat it as a problem apart, something that can be dealt with through ad hoc laws and other solutions; but the urban is certainly the most suitable dimension for a investigation of the relationships between non-European immigrants and local communities. It is in the cities that the competition for access to scarce resources is most obvious, housing above all, and the pressure exerted by the various groups is translated into explicit claims to territorial control. It is also at an urban level that the role of political mediation performed by local authorities can have the most immediate effect. The great gatherings and marches in support of or against immigrants take place in Rome, but it is in specific local situations that fires are lit and explosives used to stop social housing being assigned to immigrant workers' families. And it is in the cities that territorial conflicts proliferate (from the designation of campsites for gypsies to the use of empty buildings for the temporary accommodation of refugees).

The contradiction between the relatively small number of immigrants and the furious, often extreme reaction that their presence sometimes arouses may perhaps be explained by the fact that the entrance of this minority onto the urban scene is taking place at a time when the prevailing attitude to the processes of urban regeneration is undergoing radical change. As scholars in other countries have clearly explained with a wealth of unequivocal examples (Smith, 1996) the succession of phases of investment, dis-investment and reinvestment is not a new phenomenon, but whereas in the past there were frequent forms of social opposition to gentrification, now local gentrification has become a public politics.

With the shift of capital from one place to another, moving the frontier of profitability and causing the loss of industrial labor opportunities appears to many to be inevitable and definitive, yet urban renaissance is being advertised and sold as bringing benefits to everyone and the capacity of marginal groups to refashion space of social control into sites of resistance is diminishing all the time.

A similar tendency is also manifest in Italy, where the past offers many examples of political parties, trade unions and inhabitants engaging in joint action against regeneration plans that were perceived to be a means of expelling economically weaker groups of citizens from the more desirable urban areas. The economic anxiety and economic distress caused by forces outside their own control have placed new stresses on families and communities who in turn have changed their attitude towards minorities, thus confirming that the degree of support for social policies to address social inequality is in no small measure related to feelings of insecurity.

No-one dares any longer to oppose urban regeneration projects and anyone that impedes or slows down their implementation is seen as an enemy to be expelled or at least kept out of the way (if cities must court and compete for investors they will certainly not list the presence of "dangerous or otherwise irksome" minorities amongst their competitive advantages).

With the increasing fragmentation of society and parallel to the relaxation of state controls and to the shift in the approach of local authorities, from controlling to enabling, some have attempted to unite local people around anger at the government and minorities. This political message seems plausible to many taxpayers and so it is no longer a point of discussion whether action should be directed at people or at places. Focus is now exclusively on places, not however to improve them because poor neighborhoods make their residents poorer, but simply to promote their physical regeneration, independently of the destiny of their inhabitants; indeed such action sometimes works against the interests of the inhabitants. The recent chorus of praise with which Italians, including architects and intellectuals, have greeted the news that the British government intends to completely demolish problem districts, says a lot about the temptation to accelerate the human renewal of urban areas with development potential.

Participation-Partnership-Partnering...

We must therefore use greater caution when defining urban regeneration and similar care in attributing an indiscriminately positive connotation to drive from below, by definition considered good in contrast to the obtuseness of the institutions. If it is part of the new orthodoxy to set governance against government, and to praise the real or imagined virtues of the spontaneous and flexible networks formed by those affected by "real" local problems, one cannot ignore the experience of local authorities that have offered opportunities for communities to become involved in decision making only to see them become dominated by community élites.

Equally, one should not forget that local communities can pursue objectives pursued only by the part of the population that can express them, but that they do not produce benefits for society as a whole and that the passive acceptance of these objectives by local authorities is particularly dangerous at a time when the sense of community defined as the "feeling that you belong to a group of people because you live in the same area" has been transformed into "the fact that you live in the same area because you belong to the same group".

There is a constant crescendo of rhetorical praise for the virtues of localism, which is credited with finally ensuring that local communities have control over their own destinies by organizing themselves into networks to plan and bring into being a better world without the need of authorities or governments or politics; it is important, therefore, that we reconsider what we mean by participation.

For people of my generation "participation" is one of the magic words of our youth. It meant so much more than its dictionary definition. One wanted "to take part" not just to be able to express one's own opinion but also to give a voice to those who were not represented, not only to gain advantages on one's own territory but also to stress all the unsatisfied social needs and to give them a collective response that would result in an improved quality of life for everyone. It was a slogan that encapsulated the need for direct involvement if solutions were to be found for things that didn't work: citizens would prevail over the obtuse behavior of the State bureaucracy, the inhabitants of a certain area would themselves indicate its needs and the changes it should undergo, everyone would be authors and actors making conscious choices and decisions. It was taken for granted that this would guarantee "the general good". And we should not forget the background of labor and social conflict against which such demands for participation were made. On the one hand it often happened that district councils and union bodies supported the same requests, and on the other one there was a theoretical assumption that factories and society worked in a similar way.

But as with many apparently transgressive phenomena the connotations of "participation" shifted quickly. It became institutionalized, abandoned all pretensions to changing society and was reduced to a series of movements defending privileged or interested groups. They developed a range of ways of discovering the views of local community, and communities became effective in influencing local councils, but very few care how responsive is the council to the needs of minorities. In subsequent years "participation" was transformed into "partnership", the state of being partners in business. The changed context changed its nature.

The demand for greater participation had positive effects on a society with relatively strong institutions because it obliged them to operate more openly. But at a time like the present, when public powers are crumbling, participation takes on regressive or even reactionary connotations. Participation has become a synonym for the faculty of every local community to do as it wishes, with the result that the stronger it is the more it tends to claim advantages and privileges within its own territory.

As well as reflecting a model of a self-centered and fragmented society that follows crude but explicit slogans such as "what do we get in exchange for the taxes we pay?" or "not even one lira of what we pay in taxes must be spent outside our territory", the above attitude raises the question of where exactly the confines of "our territory" lie and who can therefore claim to be "one of us".

This question is crucial because as a result of economic restructuring "people are seeking compensation from the places in which they live" (Sennett, 1995), they are desperately seeking in the residential community "the sense of cohesion and stability that they lost in the workplace". In this process communities become defensive refuges against a hostile world and the notion of community assumes a reactionary connotation.

Press Clippings

When recounting a specific episode in which one has taken an active part and of which one has a detailed knowledge, one runs the risk of drawing arbitrary, generalized conclusions; it is equally true, however, that casting about here and there, seeking the shared elements in a variety of episodes, also raises the possibility that one's conclusions will be superficial. My own approach is not free of this latter danger.

Knowing of no case of urban regeneration in which a minority might have played a positive role and being unable to refer to any specific research in the field, I have confined myself to consultation of Italian daily newspapers in 1999 (Somma, 2000). I assembled a series of press cuttings from national dailies dealing not just with cases of ordinary racism but more particularly episodes involving city authorities (for the measures that they adopt both reflect the demands and expectations of the population and at the same time contribute to the formation of public opinion) and those that feature a more or less direct territorial element. I found a relatively high number of cases but not one of them justifies even minimal optimism.

On the contrary, the local communities that organize themselves and systematically try to ensure that their views have some impact are certainly not pressing for a stop to gentrification and marginalization; what they want is "more police, more repression and more security" and local authorities respond accordingly and often give encouragement.

The cases I have found exemplify three main attitudes, for each of which there is a corresponding type of administrative response. The first and predominant feature is the obsession with safety.

There is already a substantial body of published work on the role that the fear of crime or violence plays in transforming parts of cities into places of surveillance from which threatening groups must be excluded. There is no comparison between the situation in Italy and that in the United States, but the presence of immigrants is increasingly associated with a risk against which it is necessary to organize and defend oneself. Cases reported during the year included:

- "in the area between Lodi and Pavia, each town has its own security patrol that does the rounds of the streets twice every night";
- at Acqui Terme (near Alessandria) the council has proposed "the attribution of a prize (Lire 1 million) to anyone who reports the presence of immigrants to the police";
- at San Genesio (near Pavia) the mayor secured finance to "install three gates and shut off a series of roads";

- in Florence it was proposed that a security service should be instituted "to patrol from dusk to dawn"; according to the promoters of the idea "citizens would accept an increase in property tax in order to finance this public (!) service";
- in Genoa the inhabitants of one district organized a protest march against plans to install a campsite for "travelers" in the neighborhood; children were taken on the march.

Such news items appear frequently and show that there are demands in cities large and small in many different parts of the country that may be considered part of the movement for safer cities and defensible space and which identify the presence of immigrants and "travelers" as the chief source of danger. Fortunately, there are also stories of expressions of solidarity with immigrants and interventions by anti-racism groups.

In Voghera, for example, where it was decided to organize an unofficial referendum in an attempt to block plans to establish a camp site and a reception center for "travelers", a citizens' committee immediately organized a counter-referendum.

But to try to offset the prevailing image of immigrants as involved in crime with one presenting them as a victimized minority is of limited use, also because it risks being a form of the same deeply divisive and demoralizing ideology (Brunello, 1996).

What makes the generous manifestations of solidarity an insufficient response to the question as to "whether there exist examples of involvement of minorities in actions aimed at general improvement of urban conditions" is the lack of initiatives that propose an idea of safety "for everybody" and of projects based not only on repression but also on increased opportunities for work, housing and access to services.

The second type of message that emerges from a reading of the newspapers is more ambiguous but nonetheless dangerous. The premise "the cities are deteriorating and they must be cleaned up" is presented as an incontrovertible matter of fact; but the proposed solution – let's sweep up the city – often takes disquieting forms. The example most often quoted is that of Mayor Giuliani (he enjoys immense popularity with the Italian press and has been invited to Milan, where the Mayor often cites him as a model to be emulated) and numerous municipalities are turning to legal remedies as they seek to reclaim public space from the disturbing minorities. Though apparently neutral, populist language tends to be translated into measures that have obviously been designed with an anti-immigrant function in mind, from ordinances forbidding unlicensed windscreen cleaners and parking attendants (Turin), to reminders to taxi-drivers that it is a crime to transport prostitutes (Novara). They pretend they are punishing the act or the conduct and not the status of being part of the minority.

In Treviso, the Mayor, demanded that the army keep a closer check on the border and ordered the benches in the city's main square to be removed so as to stop people sleeping on them. Apparently the rationale was "to outlaw sleeping in public" but in reality everybody knows that in the rich city of Treviso only immigrants sleep on the streets.

This type of action too, like those in the name of security, arouse the indignation of part of the citizenry, but in general this happens only when the rhetoric of civic morality gives rise to particularly odious decisions as, for example, in Milan, where, during a period of a severe winter freeze the Mayor ordered the gates and grilles of the underground railway system to be covered with sharp needles in order to prevent people leaning against them to get some heat.

And finally, in the group of actions to "clean up the city", there are the examples of recourse to laws and regulations that are usually ignored by municipal administrations but which are sometimes activated specifically to strike at a minority.

In Alessandria, the Mayor ordered the closure of a mosque located in a building in the city center on the grounds that "there are no car parks for the worshippers!" We have nothing against Muslims is the refrain in the city, "but if they want a place to pray let them find somewhere else on the private market".

Then there is a third type of attitude that may be considered the logical consequence of the first two. In essence, the message made to follow from the obvious truth that paying attention to security and banning certain forms of behavior is not enough to guarantee "a safe city to a decent citizen" takes the form of an invitation to re-appropriate the city.

Overturning the 1968 slogan "let's take over the city" which expressed a wish to re-appropriate the city on the part of social groups and citizens excluded from the decision-making process, the motive power now represents an attempt to persuade the citizens that the decay and decline of their environment is the fault of some unwelcome minority and it takes the form of a vengeful campaign against their supposed thefts of the city. Among the groups' accused of stealing the city from its rightful inhabitants, immigrants inevitably come first.

Reinforcement of this tendency comes not only from the real hardship suffered by some groups of inhabitants who find themselves abandoned in a sort of war between paupers, but also from the simultaneous emergence of situations in which privileged areas proud of their privileges take action to keep them.

Particularly appropriate here is the term "revanchist city", introduced by Neil Smith: more than simply a divided city it is a city where the winners are increasingly defensive of their privileges and increasingly vicious in defending them, and where neighborhoods with high quality public or private amenities are open only to those who are able to pay their way in.

Of the three attitudes identified as paradigmatic in the episodes reported in the press, this is certainly the most interesting from the point of view of the editors because the reference to territory is explicit, but at the same time it is the most difficult to combat because of the transversal nature of the forces it brings together. Unhomogeneous groups, often with mutually conflicting interests – low-income tenants and property speculators, unemployed people and shopkeepers, left-wing and right-wing citizens – they all find common purpose, albeit temporary, in defense of their territory. Unlike the demands for security and decent surroundings, this one occurs mainly in the big metropolitan areas, in those urban areas, which were left to rot and ignored for years on end by local administrations and which now seem ripe for development.

In Rome, tramps and immigrants occupied the Pantanella factory at the beginning of the 1990s. After being cleared, it was bought by a well-known businessman and is now being transformed into a large commercial center, which is scheduled to open in the year 2000, in time for the Jubilee. Another area due for clearance before the Jubilee Year is in the vicinity of the former Centocelle Airport, which currently accommodates around 1500 people (groups of Bosnians, Macedonians, Romanians, Moroccans and Gypsies).

Pioltello, in Milan, is a district which was the scene of workers' struggles and public demonstrations thirty years ago and for which Fiat has plans to build a large office and commercial complex. When complaining that the area had been invaded by Romanians, the inhabitants of Pioltello described the progressive decay of their district as follows: "First we had the pollution from the factories (now closed), then the noise from the airplanes diverted so as not to disturb the fashionable district of Milano 2 (owned by Berlusconi), then a high density of *mafiosi* and now, to cap it all, immigrants".

Sell Your Ethnicity!

Beside these episodes the newspapers record many rhetorical gestures and declarations where the concept of solidarity has as much to do with show and image as anything else. A classic and so far unsurpassed example is the proposal of a number of Mayors in the Salento district that the Region of Puglia should be awarded a Nobel Prize "for the welcome it gave the immigrants".

Between the two extremes, the rise of the revanchist city and the spectacularization of charity, is it in fact possible to find the occasional example where immigrants are considered not as an obstacle to regeneration but associated with actions designed to bring about improvement? The only reports that to some degree could be interpreted as positive concern the encouragement given immigrant minorities to develop trade in their ethnic goods and artifacts.

The phenomenon can be seen as the consequence of the nature of urban regeneration, one of whose peculiar characteristics is, in order to become attractive to potential investors, employers and tourism, many districts are being converted into middle class playgrounds. In this process, they say, minorities could find a niche, converting their efforts into money and jobs, developing ethnic tourism, flea markets, ethnic shops and restaurants.

In other words, to become an attraction ethnicity must become a synonym of "exotic", and minorities have to enhance their "differentness". Thus in Milan, two days after clearing three camps occupied by gypsies, the council department head responsible announced that she intended to build a huge new structure, called Nomadopoli, that "could become an enrichment of the city, a tourist attraction where Milanese can go and experience the magic of certain of their customs". In Turin too the tendency towards ethnic festivalization is encouraged by many and is suggested as the saving situation for both San Salvario and for Porta Palazzo, the two districts with the heaviest concentration of non-EU immigrants.

Porta Palazzo, a badly run-down district that surrounds a large, permanent market area, lies in a central part of Turin. It has a resident population of about 8000, 10 percent of whom are non-Italian, but it is frequented by tens of thousands of people every day, a metropolitan-scale point of reference for the various ethnic minorities, each of which tends to congregate in the same spaces: the North Africans in a small square, anglophone Africans along one pavement, Slavs along an arcade. Periods of peaceful co-existence alternate with flare-ups of conflict and the area became the subject of considerable press comments when an "anti-immigrant" mass was said in Latin. In the words of the celebrant, the mass was "a way of repossessing our own land, our traditions and our culture, a gesture of reappropriation of an area of which the legitimate inhabitants have been dispossessed". And speaking on behalf of these legitimate inhabitants he added "we do not want to lose our identity, to disappear into the melting pot of globalism where everything is the same; we have nothing against Muslims; all they have to do is convert; it is not true that all gods are equal; ours is the true one".

A careful study of the area and its inhabitants has been carried out by Cicsene, in collaboration with Turin City Council, which has proposed to the EU that the area, defined as a "reservoir of diversity", be adopted as the subject of an urban pilot project (Cicsene, 1997). Cicsene's recommendations and the activities already put in hand are varied and carefully planned, with special emphasis being placed on educational initiatives. These are designed to reduce the diffidence that characterizes relations between the various groups by encouraging the flow and exchange of information and by fostering mutual awareness and acquaintance, also through a focus on "differentness".

A second concern is that there should be mechanisms for mediation between inhabitants and immigrants and between established local shopkeepers and new arrivals; it is clear that there is no immediate solution to the conflict so sensitive efforts will need to be made to alleviate the inhabitants' sense of injustice through some kind of "compensating" actions.

It would appear to be a realistic strategy, but it is not exactly inspiring to discover that after the enunciation of a series of ambitious objectives such as "the return of beauty to this part of the city, the control of social conflicts, the exploitation of cultural resources, the provision of spaces for amenities and services, the restoration of housing, the boosting of commerce", measures for immediate implementation go no further than encouraging multi-ethnic gastronomy and promoting "street entertainment".

Not Only "who governs and who rules" Matters, but FOR WHAT

The foregoing cannot be said to be conclusive empirical research. The newspaper reports cited have not been independently checked and it cannot be taken for granted that they give an accurate account of the situation. On the other hand, they certainly cannot be ignored in that at least to some extent they represent public opinion, and they also find confirmation in the first messages that the various political forces are sending out in the run-up to the coming round of local elections.

So if we place these apparently loosely connected episodes in the context in which they occurred which in turn is part of a larger economic shift and social movement, we can make certain observations:

- public policy and private market are conspiring against minorities and urban regeneration has become part of this policy of revenge;
- when the temptation to remove people in order to release sites with commercial potential prevails over the alternative of using the value intrinsic in the site to

improve people, protecting rather than disrupting the potential, participation is more part of the problem than part of the solution. The demand for greater participation had positive effects on a society with relatively strong institutions, but its effects may be divisive and regressive in a context of abrogating social responsibility for the city. Auditing is becoming popular, but it only collects the views of certain groups in certain parts of the city at a certain moment in time.

It seems that all sections of the population are equally likely to avail themselves of the new participation opportunities, but the voice of minorities is less likely to be heard both because fewer of them are talking and because they do not talk so loudly;

- the observable fact that local communities tend "spontaneously" to clash with minorities, with violence being greater the worse the plight of the minorities, ought to lead to external action being taken, at least to remedy the more extreme aspects of the situation. But this brings us back to where we started: the lack of a political strategy to deal with the problem.

What perhaps is needed are initiatives designed to show both the already settled inhabitants and the new arrivals that there can be areas of mutual interest; this can lead to a shared project based on living together as well as on spatial organization. The challenge is to find issues and programs that concern the whole of society and reforms that benefit all the groups in need, and that they are positive and creative rather than limiting themselves to protest.

- A news report published at the end of 1999 concerned the decision to open a Turkish bath, together with a bar, a restaurant and a library, in an area of around 3000 square meters in Porta Palazzo, on a site where there had previously been public baths. The signals the news item sends out are ambiguous. In fact the Italo-Arab Association and a committee of local residents issued a joint protest, asking "in an area where there are still many people who live in accommodation without bathrooms, which is more useful: a Turkish bath reserved for the few or amenities compatible with the needs of the poorest and most needy citizens?" But no credible alternative is called for.

Of course there is no guarantee against a bad result. Residents and newcomers may reach an agreement at the expenses of the rest of the urban community, new cleavages may appear within each group, but at the moment this seems a compromise capable of mitigating the worst effects that the quest for individual success and the desperate attempt to survive produce in a fragmented society;

- and finally, if the construction of an inclusive urban policy is to be made less difficult, a way has to be found to persuade all concerned, including those who consider themselves to be a majority, that a failure to acknowledge the rights of a minority means reducing the rights of everyone.

I found two news items especially significant and instructive in this connection. The first concerns an episode in Trento, where a private school in receipt of grants from public funds declined to accept the enrolment of a Moroccan girl who refused to attend the religious instruction lesson. The second refers to a small town in Liguria, where the mayor was arrested because he issued residence permits to Kurds, who "in exchange" had to work for him without pay in his businesses (an extreme vision of the new slogan that calls for "increased productivity" on the part of the urban poor).

Both episodes were presented as cases of "racism" and not what they are fundamentally: infringements of the rights to freedom of thought and dignity of work that belong to all citizens. Without this awareness, the suggestion that ethnic or racial minorities may become the agents of "positive" urban change and that local government will rediscover its role as a champion, helping its many different groups and interests to articulate their needs, seems to me to be unrealistic.

References

Brunello, P. (ed.) (1996), *L'urbanistica del disprezzo*. Roma: Manifesto.
Cicsene, (1997), *Un mercato e i suoi rioni. Studio sull'area di Porta Palazzo.* Torino: Agami.
Sennett, R. (1995), "Something in the city", *Times Literary Supplement.* 22 September, pp. 13–15.
Smith, N. (1996), *The new urban frontier. Gentrification and the revanchist city*. London and New York: Routledge.
Somma P. (2000), "Il razzismo", in Indovina, F., Fregolent, L., and Savino M. (eds) *1950–2000 L'Italia è cambiata*. Milano: Angeli, pp. 69–73.

Chapter 10

Concluding Notes

FRANCESCO LO PICCOLO AND HUW THOMAS

By the time the book was assembled the editors no longer worked in the same institution. Consequently, we drafted individual responses to its content. Our original intention was to meld them into a single commentary, but so different (though not contradictory) were our reactions that we decided to keep them as complementary concluding thoughts, the first five pages or so drafted by Lo Piccolo, the remainder by Thomas.

Speaking of Disempowered Groups and Organizations

Contemporary planning takes place in a context shaped by power. Assuming a traditional, or better institutional, interpretation of the term "power", we can argue that most of the groups described in the stories of this book have to be considered, literally, "disempowered". At the same time their stories suggest to us different approaches to and interpretation of the word "power".

Power is dynamic. There are many sources and definition of power: the power of a good idea, the power of coalitions and alliances, the power of authority, the power of principle, the power of commitment and networking, the power of technical skills in negotiation and dispute resolution.

Planning is often used to rationalize or advance more powerful interests. Stories such as those described in this book could be recounted in a number of ways, which would be more respectful of the structures of power. For example they might be told as tales of leadership. These stories might also be told as struggles to structure an optimum public-private deal, exposing some of the limits of technical rationality. These two approaches to story telling emphasizing leadership and professional expertise, good or bad dominate planning discourse. It is not surprising that such frames of reference frequently minimize or obscure the role of community and ethnic minorities. Leadership and professionalism, in their mainstream usage, are constructs of domination (Sandercock, 1998). It is possible to break out of those frames in telling the stories here analyzed.

These stories demonstrate that the "job" of the production and treatment of knowledge which is useful/utilizable for constructing policies can be "shared" between a number of actors but is not separable. It is mistaken to separate rigidly the role of the producer from that of the utilizer of knowledge, attributing them unambiguously to two different figures of operator. While there are forms of knowledge-for-action which are produced and treated before the action is carried out and not by the actors but by operators (technicians), nevertheless the most useful form of knowledge-for-action is that produced during the carrying out of the action itself by the actors and not by "external" operators (Crosta, 1998). This is a form of "interactive" knowledge because it is elaborated by actors who interact among themselves, but above all because it is elaborated during the course of the action (and therefore produced at the same moment in which it is utilized).

Networking as an "Alternative" Form of Representation

The experiences and the theories of "advocacy planning" and, more recently of "radical planning" are, in the light of events discussed here, subject to new interpretations, beginning with this particular and unconventional point of view. A recounting of these experiences, in fact, invite us to reconsider the mechanisms of inclusion – and exclusion – regarding citizenship, as it does the principle of neutrality and impartiality of choice, making us reflect on the potentials and limits of the processes of participation within the planning process.

What has been discussed above has highlighted the potential for a redefinition of the role of the local community, and of minority groups in particular, within the planning process, taking as a starting point the relationship between leadership and networking, and forms of representation and participation.

According to the experiences analyzed by Krumholz and Clavel (1994), if leadership is usually viewed as an important factor, coalition building (and networking) among institutional actors and local groups has to be considered central to success.

But is networking just a professional or technical skill/attitude? Some planning theorists argue that networking is just a technical action concerning (professional) planners. Benveniste (1989) identifies some skills that, if developed by planners, are quite useful and effective in a professional context. In fact, in his book *Mastering the Politics of Planning* he identifies networking, conflict resolution, coalition building and resource gathering as "techniques" and "methods" for professional planners that can be useful to relate planning solutions to political acceptability.

What most planning theorists (Benveniste included) fail to consider is the way in which people (planners, stakeholders, activists, and associations) are "differentiated" in their work of networking and coalition building. Class, ethnicity and gender do not feature as structuring forces or action arena and planners/policy analysts remain undifferentiated in terms of their social and personality characteristics (McDougall, 1993). Many historians and planning theorists do not perceive them as relevant and are remarkably silent on this crucial aspect: as a result, the planning history is full of these "black holes" (Sandercock, 1998).

The different stories contained in this book tell us about failures and successes, reactions and strategies, learning and adaptation. If some of these stories can be considered "successful stories", the reason of this success is related to networking and organizational learning. The experiences described in this book shows how planning and urban transformations are related to organizational learning, how actions and change come both from the top and the bottom of organizations, and how disempowered (minority) groups use informal networking to bring new ideas and approaches to the surface.

As John Friedmann (1995) reveals, real solutions are not found that are not the fruit of a process of reciprocal learning – a slow, difficult and often conflictual process, but one which cannot be evaded. The act of informing, or better, of reciprocal learning, is to be considered of fundamental importance in the process of building coalitions and networking, and in all acts of participation, whether direct or indirect, which assume either the formation of active political groups or the presence of aware and "informed" local communities (Friedmann, 1987; 1992). Apart from the level of efficacy in carrying out the stated goals, building coalitions and networking represent a necessary route to identifying objectives, and this even more so if there are social groups present whose requirements and wishes are – by definition – not easily analyzed and interpreted.

A central role is therefore to be granted to involvement/networking, even with the objective difficulty of putting it into practice. The experiences here described demonstrate once again that building networks is a slow and difficult process requiring an expenditure of time and resources which is considerable and often greater than the results obtained: it is an "imperfect" process, for just this absence of a direct causal link between effort expended and results obtained, as well as for the lack of a guaranteed relationship between demands expressed, real demands and results. This latter is not an aspect of small importance, and brings into play the delicate question of representation and of the relationship of correspondence or non-correspondence between a group or local community and its individual components.

Speaking about organizations, Reed (1992) has individuated five major analytical frameworks which can be considered in contemporary organization theory: organizations as social systems, as negotiated orders, as structures of power and domination, as symbolic constructions and as social practices. Consequently, negotiated order theory "emphasizes the fluid, continuously emerging qualities of the organization, the changing web of interactions woven amongst the members and it suggests that order is something at which the members of the organization must constantly work. Organizations are thus viewed as complex and highly fragile social constructions of reality which are subject to the numerous temporal, spatial and situational events occurring both internally and externally" (Day and Day, 1977, p. 132). Race, ethnicity and gender do influence these processes: as Reed (1992) has argued, there are four overlapping political areas, which produce and reproduce organizational forms and processes, namely: managerial politics, class politics, sexual politics and racial politics. The modern organization is seen to lie entwined in a complex web of political processes (McDougall, 1993).

Reference to ethnic plurality weighs this issue with still more problems, as a consequence of the heterogeneity of the groups and of the lack of connection between the different individual identities and the identity of the ethnic group, which can in no way be taken for granted. Each ethnic minority cannot be considered – in cultural, social, anthropological or political terms – as a homogeneous and undifferentiated group, fossilized in time, but contains different specificities and individual characteristics, as well as several levels of integration/ non-integration which do not make the question of "who represents who" in the network building process any easier to resolve.

The legitimacy of representation in the process of networking, or better, of the loyalty of individual representation in expressing the needs, desires and expectations of the community or of the group, remains a problem which is, in large part, unresolved. Once again, if this is a problematic aspect for itself, aside from the plurality of the ethnic group in question, then a further degree of complexity can be ascribed to the presence of ethnic plurality, there being in this case less mediation using traditional forms of political representation, particularly with reference to the objective of involving the minorities themselves. It is maintained nevertheless that through exactly this involvement of minorities in the acts of transformation and governing a city it is possible to guarantee – beyond the traditional forms of political representation – a recognition of the rights to citizenship which are often denied on a political level, and almost always on a practical level. In other words the hypothesis which emerges from the experiences here described is one of networking referred to urban regeneration and governance as an "alternative" form of representation on the part of the minorities discriminated against or excluded, or anyway unable to gain access to the traditional forms of representation. This is easier to talk about than to carry out, not only because of the evident resistance generally expressed by the majority, but also because of the difficulties, which are intrinsic to an "alternative" form of representation.

The Community Actions as Expression of the Identity and Political Role of the Local Community

In many cases that which must be viewed positively, regarding the contents and modalities of the communities actions, is the level of maturity and awareness expressed by the local communities in the formulation of a way of acting which is very complex and articulated but which isn't limited to the more elementary and undoubtedly more pressing needs, such as that for housing, but has identified "quality" objectives and strategies. This should not be taken for granted when one considers the conditions of marginality and decay in which the same communities live.

A platform of political objectives is necessary towards this end, to win consensus and to direct the conflictual energies present within the community towards common objectives. With particular reference to the realities of ethnic minorities, it is necessary to recognize that possibility for change depends principally on the capacity of the community to organize itself to make its voice heard and put forward its own requests, conquering positions of power (also) within the planning process. Communities that are fragmented, resigned or uncertain must fight against strong interests, determined to pursue their own goals. It is not possible to do so without attempting to invert this relationship of power.

According to a metaphor of Vittorio Foa (1991), we can say that there are two models of action, both in politics and in life: the model of the Castle, which moves in a straight line, through confrontation and struggle in a predefined field that cannot be avoided, and the model of the Knight, which moves sideways, searching for different fields and levels. It is possible to use the Foa metaphor to analyze the different strategies adopted by the communities, the reasons for their choices, and the way they have been forced or conditioned by all the other players and the local context. We focus on their "willingness" to involve themselves in the transformation and decision-making process, in reference to the many other different actors: public institutions, developers, investors, non profit organizations, advocates, activists, consultants.

At first sight, communities seem to have adopted completely different strategies. But at a closer inspection, if we analyze this process in its development and also consider the different possible levels of action (and time scale), these experiences suggest that it is possible to intertwine and overlap the two models. In different forms, these disempowered communities have been able to use both at the same time, in alternative ways, placing the emphasis on autonomy, local self-reliance, participation and social learning through experience.

The Utility of Some Key Concepts

The rich case studies in this book contain material that defies a simple concluding chapter. In this section we concentrate on two points. First, we contend that the case studies bear out the optimism of the book's opening chapter in relation to two explanatory notions – the social construction of minorities and social networks. Secondly, we will try to highlight a limited number of practical lessons, which can be drawn from the case studies, for those engaged in the practice of urban regeneration.[1]

In relation to the first point we have only a limited amount to add to the discussion of the book's opening chapter. The implications of a social constructionist approach to understanding minorities emerge strongly from a number of the case studies and, especially, from reading the case studies together. One implication is the need to recognize the significance of history and context in shaping social construction. The racialized distinction between Swedes and non-Swedes, for example, which Khakee and Kullander emphasize in chapter seven, would not map easily onto the kinds of distinctions drawn in Palermo (chapter six). Meanwhile, Henu, in chapter two, illustrates the significance of local context and history in influencing whether (and how) groups come to view themselves as a racialized minority (i.e. whether certain kinds of racialized categories come to have a salience, as opposed to some other kind of category, such as age). The complex case of New Orleans (chapter five) illustrates the folly of public policy attempting to import general definitions and nostrums relating to minorities into distinctive local circumstances. The nature of power – its scope, how it is exercised, and the social relations in which it is implicated – is intimately bound up with, and helps define, spatial scale; the kinds of minorities involved in explanations of power relations at any given spatial scale may not be translatable to other spatial scales, scales defined by other kinds of social relations. Moreover, not all exercises of power need be regressive, and perhaps sometimes those who find themselves at the wrong end of power relations at certain scales need not be objects of concern for those whose primary goal is a more just society – for justice requires the exercise of power. In undertaking urban regeneration, therefore, public policy must simultaneously operate at spatial scales and governmental levels where different constructions of minority-majority relations may be played out and contested. This suggests that public policy will need to be embedded in a sophisticated understanding of power relations which allows for decisions to be made about when it is more important to operate at one kind of spatial scale (and address power relations bound up with it) rather than another.

A second idea promoted in the opening chapter of the book was that of social networks and networking, as part of any helpful exploration of when and how urban regeneration activities might benefit particular groups. All of the case studies, to a greater or lesser degree, used the notion of social networks to help explain the story they had to tell. Yet there was a great deal of caution about recommending that minorities insinuate themselves into networks involved in distributing some of the good things in urban life. Both Henu, in chapter two, and Lo Piccolo in chapter six, suggested that the price for acceptance into urban policy networks was an emasculation of the objectives and claims of minority "representatives", and a consequent distancing of them from those they sought to represent. The alternative, however, might be the long struggle of the outsider for relatively modest gains, as Macchi, in chapter three, documents. Khakee and Kullander, in chapter seven, are more optimistic, about the benefits of attempting to penetrate policy networks, but their data, too, illustrate the tenacity with which those who benefit from existing constructions of majority/minority status will attempt to uphold power relations even while notionally extending the networks to include minorities. Silburn, in chapter eight, notes that some minority networks might be founded on a rejection of involvement with the non-minority activities; in that case, invitations to become involved, to use minority networks to mobilize involvement in governance, can be interpreted as threats to the minority network. There is no formulaic solution to this kind of problem – only dialogue and a gradual development of trust seems to offer a way forward.

Lessons for Policy and Practice

Some lessons for those engaged in urban regeneration have already been adumbrated. What others are there? First it is worth emphasizing that there *are* positive stories to tell. Some of these consist of concessions won by the conventional techniques of community activism (as in the cases of chapter seven) sometimes underpinned by the implicit threat of disorder (see chapter two, and Campbell, 1993). Important as these are they usually represent episodic gains, which do not seriously disrupt or reconstitute power relations in the longer term. However the case of Palermo (chapter six), is an example of an apparent incorporation of immigrant groups into a governing coalition. In this case, the definition of the groups as immigrant, as outsiders, has been the basis on which they have been identified as potential partners in governance by a Mayor keen to establish a political program with a socially progressive dimension. Acceptance of this identity is a pre-requisite of acceptance into governance networks in this case, and the longer term gains remain uncertain – but at least there is a prospect, and an appearance of thinking long term, of fundamentally changing social relations, not simply gaining a little extra bread today.

The Mayor's strategy requires the identification of organizations or individuals who can *represent* minority ethnic groups. "Represent", here has two equally important senses: a) channel and filter the views and perspectives of minorities within governance networks; b) help present a certain image of the Mayoral program to a variety of local and national audiences. So a degree of institutionalization of minority interests and identities seems to be important as a foundation for achieving influence (cf Brownill, et al 1999), which is the second lesson of the case studies. Requiring institutionalization of this kind may not be unreasonable, but it can distort social relations within a group on occasion, as Henu argues in relation to Marseilles (chapter two). Silburn, too, in chapter eight, notes that in Nottingham the decision as to whether to try to participate in voluntary community activity open to all has been a fraught political issue for Afro-Caribbean residents of an impoverished part of the city. Henu notes that the stigma associated with residence in a run down housing area is more significant than racialized identities. This phenomenon is not uncommon – see, e.g., chapter seven, on Rinkerby – though as Thomas (2000) has pointed out racialization still plays a part for it is co-residence with racialized minorities which gives an added and distinctive twist to the stigma associated with poverty. It is a phenomenon, which requires a sophisticated political and professional response, and naive distinctions based on assumed ethnicities, and unexamined pieties about racism, may hinder the development of effective community mobilization and a progressive political strategy.

It is against this background that Macchi's analysis of the struggles around the re-use and redevelopment of a psychiatric hospital in Rome, in chapter three, is so enlightening. She shows how minority identity was not something appropriated or invented in order to be used in episodic struggles, but was developed within carefully thought out, analytically self-conscious, struggles during which patient-doctor relations were questioned and re-constituted. The re-use and possible redevelopment of the hospital site has been a long running political issue, and it is surely plausible to suggest that the ability to sustain effective grass roots involvement in these circumstances, and to permeate at least some of the networks of governance, has depended crucially upon the robustness of the political analysis underlying the mobilization of the hitherto excluded. It is this need for political analysis developed by all those engaged in the struggle for justice which is the final lesson we take from the book's material.

Note

1   Some of these ideas were included in a paper delivered at the Congress of the Association of European Schools of Planning, Brno, summer 2000. We are grateful to those who commented so constructively on that occasion.

References

Benveniste, G. (1989), *Mastering the Politics of Planning*, San Francisco: Jossey-Bass.
Brownill, S. et al (1999), "Patterns of inclusion and exclusion: ethnic minorities and urban development corporations", in Stoker, G. (ed.), *The New Politics of Local Government*, Basingstoke: Macmillan.
Campbell, B. (1993), *Goliath. Britain's Dangerous Places*, London: Methuen.
Crosta, P. L. (1998), *Politiche. Quale conoscenza per l'azione territoriali*, Milano: Franco Angeli.

Day, R. A. and Day, J. V. (1977), "A Review of the Current State of Negotiated Order Theory: an Appreciation and Critique", *Sociological Quarterly*, 18, pp. 126–142.

Foa, V. (1991), *Il cavallo e la torre*, Torino: Einaudi.

Friedmann, J. (1987), *Planning in the Public Domain: From Knowledge to Action*, Princeton, NJ: Princeton University Press.

Friedmann, J. (1992), *Empowerment. The Politics of Alternative Development*, Cambridge, US and Oxford UK: Blackwell.

Friedmann, J. (1995), "Migrants, civil society and the new Europe: the challenge for planners", *European Planning Studies*, 3(3), pp. 275–285.

Krumholz, N. and Clavel, P. (1994), *Reinventing Cities. Equity Planners Tell Their Stories*, Philadelphia: Temple University Press.

McDougall, G. (1993), "Machiavelli Without the Prince", *Planning Theory*, 9, pp. 25–33.

Reed, M. I. (1992), *The Sociology of Organizations: Themes, Perspectives and Prospects*, London: Harvester Wheatsheaf.

Sandercock, L. (1998), *Making the Invisible Visible. A Multicultural Planning History*, Berkeley and Los Angeles: University of California Press.

Thomas, H (2000), *Race and Planning: the UK experience*, London: UCL Press.

Index

Printed and bound by CPI Group (UK) Ltd, Croydon, CR0 4YY

22/10/2024

01777625-0008